for *And Finally*

"Another masterful memoir from Marsh. He reviews his life, finding much to applaud but plenty of regrets, and he capably describes his experiences as a patient. The author offers a fascinating account of his often-disagreeable treatment but remains entranced by the wonders of the natural world, science, and love for his family. The conclusion finds him still alive and, readers will hope, writing another book." —*Kirkus Reviews*

"In this immersive memoir, retired neurosurgeon Marsh recalls the transformation he made from doctor to patient when he was diagnosed with advanced prostate cancer. Throughout, Marsh interweaves tender moments from his personal life, including story times with his granddaughters, with discussions of gene editing and other medical topics. Readers will find much to appreciate in this pensive probe into what it means to face mortality." —*Publishers Weekly*

"Marsh's exploration is intimate, insightful, witty, and deeply moving. . . . Marsh's writing style is such that one has the feeling of trailing behind him as an acolyte in the operating room, or in his woodworking shop, or at his dining table; in doing so, one overhears the musings of a savant, a neuroscientist, a neurosurgeon, and the inner dialogue of a patient feeling his vulnerability. He weaves in science, philosophy, history, and personal anecdotes as he tackles issues such as the nature of consciousness. . . . With this book he has left readers of the future a work to savor and learn from."

—Abraham Verghese, *The Washington Post,* and author of *Cutting for Stone*

"There's prose that breaks in gentle waves, its undercurrents deep, the surface of an ocean vast enough to put our lives in moral perspective. The narrative takes detours through DIY and dollhouses, hospital decor and Himalayan hikes. Marsh is seated, storytelling, and he is in no hurry."

—Kieran Setiya, *The New York Times,* and author of *Life Is Hard: How Philosophy Can Help Us Find Our Way*

"In the contemplation of death, Marsh illuminates the gift of life, rendering it even more precious. *And Finally* has all the candor, elegance, and revelation we've come to expect from Marsh. I read it straight through, carried along by the force of its prose and the beauty of its ideas. It's a book to treasure and reread; I'm very grateful for it."

—Gavin Francis, author of *Adventures in Human Being* and *Shapeshifters*

"In this superb meditation on life and death, Henry Marsh tackles the matter of mortality with all his trademark wit, wisdom, grace, and humility. He turns his formidable intellect and scalpel-sharp prose on himself as well as the medical profession—with marvelous results. Unflinching, profound, and deeply humane, *And Finally* is magnificent."

—Rachel Clarke, author of *Dear Life*

"*And Finally* is a close and courageous look at the prospect of death by someone who has seen it more clearly and more often than most of us, and who writes with great fluency and grace. Henry Marsh is a great neurosurgeon: he is also a very fine writer. I admire this book enormously."

—Philip Pullman, author of His Dark Materials

"He's deeply reflective; the result is a bit like sitting in the pub with the smartest person you know."

—Leyla Sanai, *The Spectator*

"It is an important message from a wise and warm narrator, and his book will bring comfort to many—and educate doctors (should any have time to read it)."

—Melanie Reid, *The Times*

"In a beautifully written memoir, the surgeon reflects on his cancer diagnosis—and explains why you should exaggerate your pain to doctors. . . . The NHS might presently be in crisis, but that is an example of the great phlegmatic British spirit we can all be proud of."

—Steven Poole, *The Telegraph*

"By sharing his findings, *And Finally* will no doubt prompt others to contemplate their own existence and, more importantly, recognize what is truly worth living for."

—*Financial Times*

Also by Henry Marsh

Do No Harm: Stories of Life, Death, and Brain Surgery

Admissions: Life as a Brain Surgeon

And Finally

Matters of Life and Death

HENRY MARSH

ST. MARTIN'S GRIFFIN
NEW YORK

Published in the United States by St. Martin's Griffin, an imprint of St. Martin's Publishing Group

www.stmartins.com

The Library of Congress has cataloged the hardcover edition as follows:

Names: Marsh, Henry, 1950– author.
Title: And finally : matters of life and death / Henry Marsh.
Description: First U.S. edition. | New York : St. Martin's Press, 2022.
Identifiers: LCCN 2022037979 | ISBN 9781250286086 (hardcover) | ISBN 9781250286093 (ebook)
Subjects: LCSH: Marsh, Henry, 1950—Health. | Neurosurgeons—Great Britain—Biography.
Classification: LCC RD592.9.M37 A3 2022 | DDC 617.4/8092 [B]—dc23/eng/20221004
LC record available at https://lccn.loc.gov/2022037979

ISBN 978-1-250-88099-4 (trade paperback)

Originally published in the United Kingdom by Jonathan Cape, an imprint of Penguin Random House UK

First St. Martin's Griffin Edition: 2024

10 9 8 7 6 5 4 3 2 1

For my granddaughters, Iris, Rosalind and Lizzie

'Two things fill the mind with ever new and increasing admiration and awe, the more often and steadily we reflect on them: the starry heavens above me and the moral law within me.'

Immanuel Kant, *Critique of Practical Reason*

'We are such stuff as dreams are made on, and our little lives are rounded with a sleep.'

William Shakespeare, *The Tempest*

I worked as a neurosurgeon for over forty years. I lived in a world filled with fear and suffering, death and cancer. Like all doctors, I had to find a balance between compassion and detachment. This was sometimes very difficult. But rarely, if ever, did I think about what it would be like when what I witnessed at work every day happened to me. This book is the story of how I became a patient.

I came to medicine by a roundabout route, which had included studying, and abandoning, philosophy. I knew only the most basic science by the time that I became a doctor. Although I am deeply fascinated by science, I am not a scientist. Most neurosurgeons are not neuroscientists – to claim that they all are would be like saying that all plumbers are metallurgists. But as I approach the end of my life, I find myself besieged by philosophical and scientific questions that suddenly seem very important – questions which in the past I had either taken for granted or ignored. This book is also the story of my attempts to understand some of these questions, without necessarily finding answers.

PART ONE
Denial

I

It seemed a bit of a joke at the time – that I should have my own brain scanned. I should have known better. I had always advised patients and friends to avoid having brain scans unless they had significant problems. You might not like what you see, I told them.

I had volunteered to take part in a study of brain scans in healthy people. I was curious to see my own brain, if only in the greyscale pixels of an MRI scan. I had spent much of my life looking at brain scans or living brains when operating, but the awe I felt as a medical student when seeing brain surgery for the first time had fallen away quite quickly once I started training as a neurosurgeon. Besides, when you are operating you do not want to distract yourself with philosophical thoughts about the profound mystery of how the physical matter of our brains generates thought and feeling, and the puzzle of how this is both conscious and unconscious. Nor do you want to be distracted by thinking about the family of the patient under your knife, waiting, desperate with anxiety, somewhere in the world outside the theatre. You need to separate yourself from these thoughts and feelings, although they are never far away. All that matters is the operating and the self-belief it requires. You live very intensely when you operate.

Perhaps I thought that seeing my own brain would confirm the fascination with neuroscience that had led me to become a neurosurgeon in the first place, and that it would fill me with a feeling of the sublime. But it was vanity. I had blithely assumed that the scan would show that I was one of the small number of older people whose brains show little sign of ageing. I can now see that although I had retired, I was still thinking like a doctor – that diseases only happened to patients, and not to doctors, that I was still quite clever and had a good memory with perfect balance and co-ordination. I ran many miles every week and lifted weights and did manly press-ups. But when I eventually looked at my brain scan, all this effort looked like King Canute trying to stop the rising tide of the sea.

Several months passed before I could bring myself to open the CD which had been sent to me after the scan. I had found plenty of excuses to put it off – downloading the data from the CD to my computer might be complicated, I had many lecture tours abroad, things to make in my workshop, days to spend with my grandchildren. In retrospect, I can see that I was apprehensive about what the scan might show, but I had managed to repress this fear, keeping it out of conscious thought.

It took only a few minutes to download the files. As I looked at the images on my computer's monitor, one by one, just as I used to look at my patients' scans, slice by slice, working up from the brain stem to the cerebral hemispheres, I was overwhelmed by a feeling of complete helplessness and despair. I thought of folk stories about people who had premonitions of attending their own funeral. I was looking at ageing in action, in black and white MRI pixels, death and dissolution foretold, and already partly achieved. My seventy-year-old brain was shrunken and withered, a worn and sad version of what it once must have been. There were also ominous white spots in the white matter, signs of ischaemic damage, small-vessel disease, known in the trade as white-matter hyperintensities – there are various names for them. They looked like some evil pox. Not to put too fine a point to it, my brain is starting to rot. I am starting to rot. It is the writing on the wall, a deadline.

I have always felt fear as well as awe when looking at the stars at night, although the poor eyesight that comes with age now makes them increasingly difficult to see. Their cold and perfect light, their incomprehensible number and remoteness, the near eternity of their lives are in such contrast to the brevity of mine. Looking at my brain scan brought the same feeling. The urge to avert my eyes was very great. I forced myself to work through the scan's images, one by one, and have never looked at them again. It is just too frightening.

There is an extensive medical literature about the white-matter changes on my brain scan, the white matter being the billions of axons – electrical wires – that connect the grey matter, the actual nerve cells. If we reach eighty years old, most of us will have these changes. Their presence is associated with an increased risk of stroke although it is unclear whether they predict dementia or not. If we make it to eighty, we have a one in six risk of developing dementia, and the risk gets greater if we live longer. It is true that a so-called 'healthy lifestyle' reduces the risk of dementia to a certain extent (some researchers suggest 30 per cent) but however carefully we live, we cannot escape the effects of ageing. We can only delay them if we are lucky. Long life is not necessarily a good thing. Perhaps we should not seek it too desperately.

I have reached that age where you start to dislike seeing yourself in photographs – I look so much older than I feel myself to be, even though getting out of bed in the morning is getting more difficult each year and I become tired more quickly than in the past. My patients were no different – they would protest that they felt so young if I pointed out to them the signs of ageing on their scans. We accept that wrinkled skin comes with age but find it hard to accept that our inner selves, our brains, are subject to similar changes. These changes are called degenerative in the radiological reports, although all this alarming adjective means is just age-related. For most of us, as we age, our brains shrink steadily, and if we live long enough, they end up resembling shrivelled walnuts, floating in a sea of cerebro-spinal fluid, confined within our skull. And yet we usually still feel that we are our true selves, albeit diminished, slow and forgetful. The problem is that our true self, our brain, has changed, and as we have changed with our brains, we have no way of knowing that we have changed. It is the old philosophical problem – when I wake in the morning, how can I be certain I am the same person today that I was yesterday? And as for ten years ago?

I always downplayed the extent of these age-related changes seen on brain scans when talking to my patients, just as I never spelled it out that with some operations you must remove part of the brain. We are all so suggestible that doctors must choose their words very carefully. It is so easy for doctors to forget how patients cling to every word, every nuance, of what we say to them. You can unwittingly precipitate all manner of psychosomatic symptoms and anxieties. I usually told cheerful white lies. 'Your brain looks very good for your age,' I would say, to the patients' delight, irrespective of what the scans showed, provided that they showed only age-related changes and nothing more sinister. The patients would leave the clinic room smiling happily and feeling much better. The eminent American cardiologist Bernard Lown has written of how important it can be to lie to patients – or at least to be much more optimistic than the facts perhaps justify. He tells stories of patients of his who were close to death from heart failure but who rallied and survived when he was overly positive.

Hope is one of the most precious drugs doctors have at their disposal. To tell somebody they have a 5 per cent hope of surviving is almost as good as telling them they have a 95 per cent chance, and a good doctor will emphasise the life-affirming 5 per cent without denying or hiding the corresponding 95 per cent probability of death. It is Pandora's Box – however many horrors and ailments come out of the box, there is always hope. Only at the very end does hope finally flicker out. Hope is not a question of statistical probability or utility. Hope is a state of mind, and states of mind are physical states in our brains, and our brains are intimately connected to our bodies (and especially to our hearts). Indeed, the idea of a disembodied brain, promoted by the more extreme protagonists for artificial intelligence, might well be meaningless. This is not to say that being kind and hopeful will cure cancer or enable us to live for ever. The human mind is always trying to reduce all events to single causes, but most diseases are the product of many different influences, and the presence or absence of hope is only one among many.

I have a fine copy of an engraving by the sixteenth-century German artist Albrecht Dürer on the wall of my study, which I inherited from my mother. It shows St Jerome at his desk in his study, a beautiful medieval room with a coffered ceiling and a large window with deep reveals and small panes of crown glass, through which diagonal sunlight falls. There is a lion – the creature associated with the saint – sleeping on the floor in front of him. The story goes that he removed a thorn from its paw, and it became a sort of pet. Next to the lion is a dog, symbolising loyalty. St Jerome was one of the early fathers of the Christian Church. He is said to have had an enthusiastic following of wealthy widows in fifth-century Rome. One of their daughters came under his influence and his advocacy of leading an ascetic life. He was blamed for her death from what some say in retrospect sounds like anorexia nervosa. I suspect I would dislike St Jerome if I met him, and consider him to be a fanatic. And yet I love the Dürer etching, with its aura of wisdom and learning, and I once made a table modelled on the desk in the picture. There is a skull on a shelf beside St Jerome's desk, an icon often found in images of medieval philosophers. *Memento mori*, a reminder of the death to come. My brain scan is no different. However hard I looked it at – and I looked very hard – it told me nothing that I did not know already: my brain is ageing, my memory is not as good as it was, I move and think more slowly, I will die.

I should have known that I might not like what my brain scan showed, just as I should have known that the symptoms of prostatism that were increasingly bothering me were just as likely to be caused by cancer as by the benign prostatic enlargement that happens in most men as they age. But I continued to think that illness happened to patients and not to doctors, even though I was now retired. Twenty months after I had my brain scanned, I was diagnosed with advanced prostate cancer. I had had typical symptoms for years, steadily getting worse, but it took me a long time before I could bring myself to ask for help. I thought I was being stoical when in reality I was being a coward. I simply couldn't believe the diagnosis at first, so deeply ingrained was my denial.

For many years I had kept a human skull on a shelf in

my study, in rather self-conscious imitation of the Dürer. I had found it discarded in a box, in a pile of rubbish, when the hospital I had worked in for many years was closed and relocated. An unknown predecessor must have been practising on it – there was a series of burr holes in the skull's vault, with saw cuts between them, that had clearly been made with an old-fashioned surgical hand saw that fell out of use a long time ago. It was otherwise in unusually pristine condition, with intact stylomastoid processes, the little needles of bone behind the ear to which the stylomastoid muscles are attached. They often break off quickly with any handling of the skull. Once I had been diagnosed with advanced cancer, I no longer found the sight of the skull amusing. I gave it to one of my colleagues at the hospital where I used to work, so that he could use it for teaching.

2

We are condemned to understand new phenomena by analogy with things with which we are already familiar. The history of our attempts to understand our brains, and in particular the relationship between its conscious and unconscious actions, is a history of metaphors.

When physicists started to explore the structure of the atom, they visualised it as a miniature planetary system with electrons orbiting the nucleus, although the nucleus was so small relative to the atom that it was compared to a fly in a cathedral. Electrons were thought to spin as they circled the nucleus, but as experiments revealed the increasingly bizarre behaviour of particles at a quantum level, this description – derived from the macroscopic world – failed. The spin of an electron has become simply the name for a mathematical property of electrons that explains experimental results with great accuracy. The trouble with neuroscience is that we have yet to find appropriate metaphors for our brains, let alone profound mathematical equations. Optimists assure us that one day we will have such equations, just as powerful as the equations of quantum mechanics, but I have my doubts. You can't cut butter with a knife made of butter, as a neuroscientist friend of mine once told me.

Despite the vast amount of neuroscientific research

that has been carried out since the seventeenth century, our understanding of our brains remains remarkably limited. We can describe, in great detail, the topography of the brain – which parts are involved in particular processes, be they thoughts, feelings or movements, and some of the underlying electrochemical physiology. I think of all the beautiful maps and diagrams of the brain that fascinated me when I was a medical student, and still do. We are like explorers discovering an ancient, abandoned city. We can describe the bricks with which the city is built, and how they are cemented together. We can draw precise maps of the city and its roads and buildings, but we have little, if any, understanding of the network of lives that were once lived there. But this is an analogy, and far from accurate.

The anatomists of old named some parts of the brain after fruit and nuts – such as the amygdala (almonds) in the basal ganglia and the olive in the brainstem. Although Hippocrates, in the fourth century BCE, had firmly placed the brain at the centre of human thought and feeling, most early medical authorities thought it was of no importance. Aristotle thought that the brain was a radiator for cooling the blood. The only important parts of the brain for Galen, 500 years later, were the fluid cavities – the ventricles – in its centre, and not the tissue of the brain itself. The early medical authorities understood the body and brain in terms of fluids – the four humours – of black and yellow bile, blood and phlegm. These fluid excretions were, after all, the only access doctors had to whatever might be going on inside their patients' bodies. With the rise of the scientific method in the seventeenth century the brain started to be described in terms of the latest modern technology. Descartes explained the brain and nerves as a series of hydraulic mechanisms. Hydraulic technology was not new at that time – Archimedes and others had used it extensively in the ancient world – but what was new were the systematic investigations of hydraulics carried out by researchers such as Galileo, and the creation of complicated fountains in the fashionable palaces of the wealthy. In the nineteenth century the brain was explained in terms of steam engines and then telephone exchanges, though Freud's psychoanalytic theory at the turn of the century was still couched largely in hydraulic terms, that made the id and ego sound like the components of a flushing toilet. And in the modern era, of course, the brain has been seen as a computer.

But we have never met our brains before, and perhaps we lack the metaphor with which to understand them. When I lie in bed in the morning, struggling to get up – a problem that has come with retirement and became much worse with hormone therapy and radiotherapy for cancer – I find it hard to escape a marine metaphor. My conscious self is like a small boat sailing on a deep ocean, or perhaps more like a submarine that comes to the surface when I awake. I then delude myself that I am steering the boat, when in fact its course is determined by the wind and the deep currents. But this is a false metaphor, of course, as my conscious and unconscious selves are part of the same phenomenon, in a way that we find impossible to describe. The submarine is part of the ocean, not separate from it. Most writers trying to describe the relationship between the conscious and unconscious sink into a muddled flood of analogies and metaphors – or perhaps I should more modestly say that I become muddled by what I read. My conscious and unconscious selves (for want of a better word) are made of the same material – the electrochemical activity of my 86 billion nerve cells. 'I' am both my conscious and unconscious – they are not separate entities.

Some psychologists and philosophers delight in telling us that our sense of self is an illusion. I briefly studied philosophy at Oxford University, but eventually fled to the more practical world of medicine. But at least I learned from studying philosophy for one year the importance of the phrase 'It all depends on what you mean by'. The word 'self' is not easy to define, and the word illusion simply means that something is different from how it appears. I have no intention of going down the rabbit hole of what the word 'self' means, but I realise that I find it very hard to know what it is that I might be losing as my brain shrinks. How can I compare myself now with my past self?

As I lay in the scanner, I wasn't worried about what the scan might show. I was more concerned about whether I would be able to keep still for an hour, with a plastic visor over my face, and headphones over my ears. In the event it was easy. I went into a trance, fascinated by the strange sounds of the machine at work.

MRI scanning is all about quantum mechanics. All I can understand about quantum mechanics is that you cannot understand it – at least not by reference to everyday life. You just have to accept that in the microscopic quantum world particles of matter behave in ways that are impossible in the everyday world we experience with our senses. They can be both waves and particles at once; they can tunnel and entangle, and be in two places at the same time, engaging in 'spooky action at a distance' as Einstein put it. How this crazy microscopic world is assembled into the macroscopic world where, unlike quantum particles, we cannot walk through walls or be in two places at the same time, is far from clear – but somehow, it is.

When you lie in an MRI scanner you are in an immensely strong magnetic field, which forces the protons in the hydrogen atoms of the water of your body to 'spin' in the same direction. The physicists tell us that this is not to be thought of as spinning like a planet or a child's toy top, but as an abstract mathematical quality. Your straightened out, magnetised protons are then bombarded with a pulse of radiofrequency radiation, the same sort of radiation as visible light but of a different frequency. It is of low energy and described as non-ionising radiation – in contrast to high-energy ionising radiation, that knocks electrons off atoms and can break the chemical bonds that keep molecules glued together and causes damage. Ionising radiation is used, for instance, with radiotherapy for cancer. As I was soon to discover.

The radiofrequency pulse in an MRI scanner knocks magnetised protons out of alignment, adding to their energy. Once the radiation has stopped, the 'excited' protons then 'relax', releasing the extra energy by emitting radiation in turn, which is picked up by receivers which are used to generate the scan images. There were crackles, rumbles and staccato explosions like machine guns. There was thunder and then soft humming, all with entirely unpredictable rhythms. The ageing, living matter of my brain was being magnetised and irradiated, but I felt nothing.

It is notoriously difficult to visualise large numbers. The hunter-gatherer Pirahã tribe in the Amazon do not count beyond three – they only have the word 'many' for any numbers greater than three, and apparently have difficulty solving simple arithmetical problems. For which, one should add, they have no need. According to Daniel Everett, the Christian missionary turned anthropologist, who learned their unique language, and lost both his faith and family in the process, they do not worry much about the future. They are quite free from the mental illnesses of western life and have no need for numbers beyond three.

The invention of microscopes and telescopes in Europe in the sixteenth century revealed new worlds – worlds of vast numbers. The unaided human eye can see only a fraction of the stars above us at night and only a fraction of the electromagnetic spectrum. We are entirely unable to see the cells of which we are made (with the exception of the egg cells in a woman's ovaries), or the bacteria and viruses that surround and inhabit us. Our increasing, but still very limited, understanding of our brains has been made possible by new technology – by the discovery of electricity, by microscopy, by brain scanning.

We start life as a single cell but end up as creatures of some 30 trillion cells, within an even greater number of bacteria, upon whom we depend, in our intestines and on our skin. Our hearts will beat, on average, about 4 billion times (and it is roughly the same for most animals – a mouse's heart beats at 500 times a minute, the long-lived Galapagos turtles at four times a minute). The earth is 4.5 billion years old, the universe is about 14 billion years old. There are almost 8 billion humans alive on Planet Earth.

I find it very hard to comprehend that 'I' am the 86 billion nerve cells of my brain. There are at least 500,000 kilometres of wiring connecting the nerve cells together at junctions called synapses – greater than the distance from the earth to the moon. A cubic millimetre of cerebral cortex (the surface layer of the brain) contains up to 100,000 nerve cells and a billion synapses. It has been estimated that there are about 125 trillion synapses in the adult human brain.

A standard building brick is sixty five millimetres thick. 125 trillion bricks (the number of synapses in our brains) stacked on top of each other would reach way beyond Pluto and the solar system. I find these figures so improbable and incomprehensible that I have to keep on checking the very simple arithmetic involved, worrying that my ageing brain can no longer deal with the decimal points and exponentials that in the past would not have troubled me. But it is also because we simply cannot envisage such colossal numbers and exponentials and can only understand them in the form of mathematical formulae. The bricks are just a metaphor. All sorts of metaphors have been used to try to help us visualise large numbers. But they are all hopeless – faced by these unimaginable numbers, we are as helpless as the Pirahã tribe, struggling with simple arithmetic. As helpless as I felt when looking at my brain, and what the scan told me.

A nerve cell (neuron) can be described as an input/output device. Nerve cells can vary greatly in their structure, but what they all have in common is an axon, a cell body and dendrites. The dendrites are the input devices and form a forest of branches that grow out closely from the cell body and are studded with synapses. The axon – the output device – is a single cable-like structure that projects from the cell body and is connected to the dendrites of other nerve cells at the synapses, and to muscles and organs in the body outside the brain. The axons can vary in length between metres for the nerves controlling the muscles of the legs, to millionths of a metre in the brain itself. Each of the 86 billion nerve cells can be connected to tens of thousands of other nerve cells at synapses. Electrical impulses travel down axons and connect to nerve cells via the dendrites. These impulses can either encourage or discourage the nerve cells to fire an impulse down their own axons onto other nerve cells. Whether the nerve cells fire or not will depend on the net effect of the thousands of messages their dendrites are receiving from other nerve cells.

86 billion nerve cells, 125 trillion synapses. But it gets even more complicated. Nerve cells are not simple on/off switches, but vary the rate at which they fire, and can send long or short trains of impulses down their axons. It is tempting to assume that the pattern of the nerve cells' firing is a code – in principle no different from Morse code – but in reality we have no idea whether this is the case or not. At the moment it's a question of faith, not evidence, that the brain computes. This probably explains why arguments about whether brains are like computers or not can become so heated.

Recent research also suggests that the 125 trillion synapses are not simple switches that are entirely controlled by the electrical impulses they receive, and that they have a degree of independence. Nor is the connection of the axons to the dendrites a direct electrical junction (although they also exist in the brain). The axonal part of a synapse releases chemicals – neurotransmitters – that act on the dendritic part and alter the receiving cell's electrical state and so pass on the nerve impulse. When I was a medical student in the 1970s, only two neurotransmitters were known – noradrenaline and acetylcholine. Now at least a hundred have been identified.

And then there are at least a further 85 billion glial cells, packed around the nerve cells. These were once seen as little more than polystyrene packaging, just as much of the DNA in our chromosomes was once dismissed as unimportant 'junk'. Both assumptions have proved to be wrong.

And out of this extraordinary, unimaginably complex and circular dance of nerve cells stimulating and inhibiting each other, reacting to the outside world and to the body of which the brain is part, thought and feeling, colour and sound, pain and pleasure, all arise. And my feeling of being me, and my dismay at seeing my brain scan. And we have no explanation whatsoever as to how all these different experiences arise from the same physical processes.

My brain is now shrinking. It is not yet known how much of this is due to shrinkage of my brain's white matter – the myelin-insulated axons that connect nerve cells – and how much to the death of the actual nerve cells themselves, which form the so-called grey matter. But it would be a mistake to think that your mental abilities are only a matter of the number of cells and synapses in your brain. In its first eighteen months a baby has many more synapses in its brain than an adult. The development from then on is as much about removing synapses – 'synaptic pruning' – as it is about creating new ones. The brain is sculpted by experience, cutting away connections that are not being used. Until the age of two, for instance, children in all cultures can discriminate between all the basic sounds of *all* languages, but from then on, they can only recognise the phonemes of their mother tongue. Chinese children, for instance, lose the ability to distinguish between the consonants l and r. And we know that early experiences in childhood – especially of deprivation – have catastrophic effects on later life. Perhaps I am reading too much into the decay revealed by my brain scan. Perhaps I should find some consolation in the thought that mental abilities are not just a matter of brain size and the number of synapses. I love watching my granddaughters rushing about, shouting, playing, full of life – they have an extraordinary and wonderful ability to learn. As I struggle with the maths lessons I have with my neighbour – a very good friend and retired maths teacher – I envy them a little.

3

I started keeping a daily diary when I was twelve years old – a consequence of the increasing self-awareness and self-consciousness that comes with testosterone and puberty. I destroyed it all ten years later, as I felt so embarrassed by it. I regret this now – it would be interesting to have some insight into the awkward child that once I was. Our memory of our past self is largely a self-serving construct, with only a few certain facts thrown in, and a diary can provide some objectivity. I have continued to write almost every day – it has become both a compulsion and a duty – and yet rarely look at it. But I did consult it to find out what I had been thinking at the onset of the Covid-19 pandemic.

Not until 23 February, exactly one month before the lockdown started in England, did I mention coronavirus in my diary, although we had all heard many weeks earlier about what was happening in a place called Wuhan. Wuhan could have been a small village for all I knew – in fact it is a city with a population of 11 million. But Wuhan was far away and of no concern to us. I wrote that the coronavirus appeared to be spreading and I thought that it might be serious and a threat to my wife Kate, as she is on immune-suppressant drugs for Crohn's disease. And yet, as far as I can remember, I did not feel particularly anxious. My diary made no more mention of it for the next few days, and instead continued with its usual reporting of the weather (continuous rain), my work renovating the derelict lock-keeper's cottage beside the Oxford canal that I had bought some years earlier (progressing very slowly), and my regular exercises (painful and tedious but occasionally exhilarating). Four days later, I commented that the news was all about coronavirus, that I was worried again about Kate and the possibility of her dying; I then contradicted myself by adding that I was not sure whether the virus was a major threat or not. It seems I was more concerned about whether the floor joists in my flat in Oxford would take the weight of a bookcase I was building.

I was planning on selling my house in London in a year's time and moving, a little reluctantly, to Oxford. I was hesitant because I would never be able to accommodate in Oxford all the books and possessions, tools and wood that I had accumulated in London. I had bought the derelict lock-keeper's cottage, shortly before retiring, with a view to having a workshop in the garden, and a rural retreat, but as the cottage was very small I would still have to get rid of many possessions. I love my house in London. I have spent twenty years working on it, improving it, with oak floorboards and fanciful overdoors, a loft conversion and bookcases in almost every room. And there is also the garden – a small and unruly paradise, where I keep my bees, with a workshop at the end, where I keep my many woodworking tools. But I have also had to confront the fact that much of the work I have done on the property is of poor quality and needs to be redone. And there was also the disquieting thought – what was the ultimate purpose of all that frenetic activity and accumulation of possessions in the first place?

At the end of February, I noted that the coronavirus would clearly become a pandemic. I wrote that I was not concerned for myself but feared that Kate would not survive. I went to bed that night with apocalyptic thoughts about the pandemic. A medical acquaintance of ours with a high public profile had fallen ill with the virus and was badly affected. She emailed me telling us to wear gloves when opening the post. She appeared – looking haggard – on the BBC news, saying how awful she had felt. This made me angry – I felt she was panic-mongering and overreacting. It is embarrassingly obvious to me now that this was denial on my part. I was still finding it very difficult to accept just how great were the changes that were coming. When faced by uncertainty, we all oscillate wildly at first between panic and dismissal – as I was to discover when my cancer was diagnosed.

Over the next three weeks Covid-19 started to take over our lives. There was excitement – of a sort – when the lockdown was eventually imposed on 23 March. With neither cars nor aeroplanes the streets and sky suddenly fell silent, and the shelves in the shops were suddenly stripped bare. Shopping became something of an adventure, with the fear that going out you might catch the virus, and the difficulty of finding something to eat. Loo paper disappeared so I took to perching over the bathroom basin. The news reported ever-increasing deaths, especially in the elderly, in which category I now reluctantly had to include myself. I wore gloves when going out and washed whatever I had bought with dilute bleach.

When the government appealed to recently retired doctors to return to work, I did not hesitate to volunteer, liking the idea of heroic self-sacrifice, of answering the call of duty and wanting to be important once again, if only in a little way. Just as with my early thoughts about the pandemic, these feelings were somewhat abstract. But as I worked through the bureaucracy involved in returning to work, I started to become very anxious. Hospital staff were falling sick in large numbers, and some were dying. Feeling cowardly and a little hypocritical, I told the people organising the return to work that I did not think that, at my age, I should be working on the Covid wards. On the other hand, the utility of an elderly brain surgeon in the emergency was not at all clear, either to me or to the NHS.

In the meantime, Kate had fallen ill with a chronic cough and fever. She was in Oxford and I was in London and it sounded as though she had the dreaded virus. The chaos of the government's muddled and delayed response to the pandemic meant, of course, that she could not be tested, but we agreed that I should remain in London, for fear of her infecting me. I spent the next few days overwhelmed by my love for her, fearing that she might die. This was not an exaggerated fear – at the time of our marriage sixteen years earlier she had developed severe pneumonia after going down with the flu. She hates hospitals, so I had treated her at home with antibiotics that I arranged through my juniors at work. This was not the first time she had come close to dying – she had once suffered from intestinal obstruction and an abscess caused by her Crohn's disease.

I played a series of scenarios in my head of her death from coronavirus, all of them very distressing, at the same time thinking of our past and happy twenty years together. Psychiatrists might call this catastrophising on my part, and perhaps I was exaggerating, but having prepared myself for the worst, I found it easier to put the problem to one side and await events. We spoke to each other on the telephone several times a day, and each time Kate – coughing and croaking – would assure me that she was going to get better. And as the weeks passed, she did indeed slowly improve.

The lockdown was an entirely novel and universal experience, overshadowed at first for me by Kate's illness and the constant stream of death statistics in the media, as though there was some kind of deadly Olympiad in progress, with countries competing for the most deaths. I was also frightened that I would fall ill as well. But as it became apparent that a dry cough and fever were only two of the ways that Covid-19 could present, I realised that I had very probably contracted the infection myself several weeks before the lockdown. I had been to Ukraine in early February – a country where I have been working *pro bono* for many years – and lectured at the medical school in the city of Ivano-Frankivsk. Shortly after my return I developed severe abdominal pain, quite unlike anything I had ever had before, and rigors – the medical name for shivering attacks. I retired to bed and took my temperature, which to my surprise was normal. A few days later I developed a sudden nosebleed – the first one for over fifty years. I was disconcerted but thought nothing of it. When Kate had much more typical symptoms of Covid infection, and also a nosebleed, I put two and two together. I found a medical article on the Internet describing rare abdominal presentations of Covid infection. So I concluded that I had probably contracted it and was probably now immune. This was far from certain and may well have been wishful thinking, but I chose to believe it nevertheless, and it was a great relief.

I would never have guessed at the intensity of what I felt once I had decided that I was probably immune to the virus – feelings not of fear, but of remission, of brief reprieve, almost of sanctuary, and also of great loss.

The lockdown coincided with perfect spring weather – so fine, prolonged and warm that it spoke of climate change. The bushes in the little paradise of my back garden almost all burst into flower all at once, and the trees went from being bare winter skeletons to towers of spreading green leaves in a matter of days. The bees came rushing out of their hive in front of my workshop and shot up into the sunlight, rejoicing in vertical zigzags.

And the lockdown brought complete peace and quiet. The air felt as fresh as if you were in the countryside and the sky was a clear and deep blue. The only sounds were of birds singing, children playing and the wind in the trees. And at night, at first, there was a full moon, looking down kindly on the suddenly silent city, and you could see the stars. It was a vision of heaven, here in London, SW19. Time had stopped. Eternity is not the infinite prolongation of time but instead its abolition. (Cosmologists tell us that time really can stop – but at the event horizon of a black hole rather than in a back garden at the bottom of Wimbledon Hill.) In the evenings I would sit in my garden and look up at the tall trees in the little park beyond my workshop, illuminated by the setting sun against the infinite, deep blue sky and for a while feel quite swept away.

But I also felt an almost overwhelming, tragic sense of loss. Life had come to a complete stop, and so you could see clearly into both the past and the future, no longer blurred by incessant movement. My home, in which I have invested so much time and energy, became a place of great beauty and sadness – I will have little choice other than to leave it, as a first step towards old age, debility and then my death. The silence and clear air, and the return of birdsong, reminded us of what we have already lost with cars, pollution and the changing climate, and the unnaturally fine weather told us that Nature is out of joint, and that there is much, much worse to come.

As I looked at the trees in the wind, I was well aware

of how lucky I was compared to most people during the lockdown, with my hand-built home and garden and workshop, and pension and good health. I had no idea that soon that would change for ever. And I thought of the soulless, treeless Soviet suburb of Trojeschina in Kyiv, where I used to stay when working with my colleague there. What must it be like to grow up without trees.

4

In retirement, memories of my former patients often suddenly appear, like illuminated capitals in a book of plain, technical text. Looking at the trees, I suddenly remembered a man from Ecuador with a brain tumour. He was a botanist who worked in the tropical rainforests. His sister lived in London, I think, which is why he came to see me. Once he was back at work, he sent me a letter with some photographs of the rainforest, telling me of his deep love for it. I remember being very moved – I wish I had kept the letter. The tumour was ultimately incurable, and he died some years later – but not before his sister had written to me despairingly, asking for help, which I could not give. I remember him, and my ultimate helplessness, so clearly that it hurts.

After I had been diagnosed with advanced prostate cancer and become a patient myself, I was surprised to keep on remembering more and more patients whom I had completely forgotten – some were cases going back more than thirty years. Now that I was so anxious and unhappy, feeling abandoned, I realised how anxious and unhappy so many of my patients must have been, and yet how I had chosen to turn a blind eye to this. My former patients became reproachful ghosts who came to punish me. They were everywhere, lurking behind everyday thoughts and sights and sounds. I thought of how I would be a much better doctor if I could start all over again. How I would be full of the compassion and understanding that I lacked when I was younger. But would this then leave me unable to enter the operating theatre, pull on my gloves, and cut into my patients' heads? I simply don't know, although I do know that the arthritis I am starting to suffer from in my hands makes it all a nonsense anyway.

As a doctor, you could not do the work if you were truly empathic – if you literally felt yourself what your patient was feeling. Empathy, like exercise, is hard work, and it is normal and natural to avoid it. You have to practise instead a limited form of compassion, without losing your humanity in the process. While I was still working, I thought that I had achieved it, but now, looking back, and as a patient myself, I was full of doubt.

A much greater sin than detachment is complacency. If you are detached you can, at least in principle, still do the work well, even though it will involve feigning concern for your patients. I know a few surgeons who did not seem to suffer much when their patients came to harm, but they are very much in a minority. If you become complacent, however, patients will suffer, as you will accept bad results, and stop trying to do better. It is remarkable at departmental and multi-disciplinary meetings how easily complacency and 'group think' sets in. Nobody wants to rock the boat, to be a difficult colleague, and mistakes and bad results are quickly brushed under the carpet. It is the very opposite of what such meetings are supposed to achieve. And in the world of modern 'managed' healthcare, where doctors have little of the autonomy I had years ago, it is all too easy to blame lack of resources for bad results, and say they are somebody else's fault, and do nothing about them. But to be honest, even when doctors had much more independence and autonomy than they have now, it was still very easy to become complacent. All our patients die eventually, and it is a question of judgement – with all the inconsistencies, frailties and biases that that word involves – whether a death was avoidable or not.

After retiring I continued to attend occasional Morbidity and Mortality meetings in my former department, in my role as wise old man. I found it very difficult. I was now a patient myself, with a brain scan of my own, showing ominous changes. I watched the parade of patients' scans projected onto the wall in front of us – images of diseased and injured brains and spines, many with cancer. As I listened to the junior doctors stumbling through their presentations of the human tragedies behind each scan, with the seniors in attendance usually silent and seemingly uninterested, I found myself torn in half.

The simplest way of limiting our empathy and compassion is to divide the human race (and all living creatures) into 'Us and Them'. Research shows that we start doing this when we are only a few months old. When I was still a pre-clinical medical student, and not yet on the hospital wards, I was distressed when some patients were brought to the medical school for a demonstration. We sat in the lecture theatre – an old one, with a tiered semicircle of wooden benches – and looked down on a procession of patients, who were brought in one by one. The demonstrator was a neurologist from the famous neurological hospital half a mile away. He took great delight in demonstrating the patients' 'signs'. One of the patients was a young man with an inoperable spinal tumour.

'Used to be a strapping young soldier,' the demonstrator said with relish, as he told the poor man to strip to the waist, and then showed us the profound weakness of his muscles and picked up a reflex hammer with a flourish and – to use the medical language – elicited his 'pathologically brisk' reflexes. Neurologists have always set great store on showmanship when demonstrating patients' signs. Like conjurers, they would pull the diagnosis like a rabbit out of a hat, until brain scanners were invented and showed that not infrequently, they were wrong. Perhaps the demonstration was made all the more distressing by the fact that the patients came in fully dressed, and were made to undress in front of us, and this was not in a hospital. Once in a hospital, patients become instantly dehumanised, and it is much easier to view them dispassionately.

This was almost fifty years ago, and things are better now. I served as an examiner for the Royal College of Surgeons for many years for the final FRCS examination, part of which involved patients. We bent over backwards to treat the patients respectfully. As it was, I think many of the patients quite enjoyed the experience of seeing the anxious examinees grilled by the examiners. It was a role reversal of sorts, with the usually suave and self-confident young doctors frightened of the patients, instead of the other way around.

Shortly after I was diagnosed with cancer, I cannot remember exactly when, I had a very vivid and intense dream. The old family dog, whom I loved dearly but had once tormented when I was young, came stiffly towards me. He was old and grizzled and arthritic, and I stroked his head, saying – or thinking, I'm not sure which – 'Now we are old together, and falling apart. Come on, I'll let you out into the garden so you can empty your bladder – we both have the same problem now. And we are both near death.' The dream brought an intense feeling of love and reconciliation, although it was only myself who was doing the forgiving. I had gone to bed anxious and unhappy but woke from this dream refreshed and happy, and completely at peace. I got up and went for a five-mile run. It was the first such run I had actually enjoyed for many months.

The memories of former patients were to become less intrusive over time, but I don't know if this was because of the dream. I had wondered, when memories of former patients kept on appearing, if part of me hoped that by acknowledging that I had neglected them, I would be saved. A confession, a form of magic, a fairy story.

5

After I retired from full-time surgical work at the age of sixty-five, I continued to work abroad, mainly in Nepal and Ukraine, until Covid-19 took over the planet. In Nepal I would put in a few weeks in the hospital run by my colleague and friend Dev (more properly called Professor Upendra Devkota) and my son William would then come out to join me and we would go trekking in the Himalayas.

On our last trek, in December 2019, we probably went up too quickly, ascending 2,000 metres in two days to Gosainkunda at 4,300 metres. We had become a little overconfident from previous treks, when we had gone higher without any difficulties, but had climbed more slowly. Nor, on this trek, had we bothered to take the drug acetazolamide, which can help with mountain sickness.

Gosainkunda is a rather bleak Alpine lake high in the Himalayas – an important place in Hindu mythology, and a place of pilgrimage. The blue-skinned god Shiva, his throat burning from poison (self-administered to protect the three worlds, it is said), struck the ground with his trident in the search for water, thus creating the River Trishuli. (In reality, the river arises to the north, in Tibet.) Apparently, if you bathe in the lake, it will wash away your sins, but I didn't try – it was far too cold. The lake is surrounded by low mountains. The snows had not yet come and the mountains were rust-coloured with sea buckthorn – a short shrub, which produces small and bitter berries that are claimed to cure all manner of ailments and balance your immune system, whatever that might mean. I didn't try them. The river rushes down from the mountains, the water blue and white, a glorious glacial river, bounding enthusiastically over granite boulders, heading south across Nepal. On previous visits to Nepal, when travelling with Dev to outlying hospitals, we had often stopped at a small resort on the road from Kathmandu to the Indian border. It was on the side of the valley, high above the Trishuli, and we would sit there drinking coffee, admiring the river below us – now slow and wide – and the intensely green hills on either side, with serried terraces of rice paddies. Dev would talk of his childhood, in the days when the only bridges across the river were made of rope and there was no road from India. Porters had to travel by foot to India, returning with sacks of salt on their shoulders. Although there is now a road, it is notoriously dangerous, with buses regularly falling over its edge into the ravine below, killing many people. I personally saw the immediate aftermath of two such accidents – a silent crowd standing on the roadside, looking down at the smashed-up bus hundreds of feet below.

Dev died a year before my son and I saw Gosainkunda. He died from what the obituaries call a short illness – from cholangiocarcinoma, a rapid and invariably fatal cancer of the biliary tract in the liver. He came to London for treatment and spent almost six months in hospital before it became clear the tumour was progressing despite chemotherapy. His wife was with him and asked me to visit. By this stage he knew that he was dying, and he wanted to say goodbye to me. I cannot remember the words we used but when I got up to leave, he drew me to him, and we embraced. When my time comes, I thought, as I walked away down the long hospital corridor outside his room, I hope I manage to be equally dignified. With great difficulty, he was able to get back to Nepal two weeks later and died in his own hospital. I miss him greatly.

The Trishuli winds its way down to the Terai – the lowlands of Nepal, which were once infested with malaria until they were sprayed with DDT in the 1960s. Until then the only people who could live in the Terai were the Tharu people, who had evolved natural immunity. After the eradication of malaria, millions of southerners from India migrated north into the Terai, causing ethnic tensions that have continued ever since. The river – now flat, slow and much wider – then flows into India, where it eventually joins the Ganges. By the time it reaches the Bay of Bengal, hundreds of miles later, it is dead – full of toxic rubbish, poisoned and polluted, like so many of the world's great rivers. A year earlier I had been lecturing in Karachi. I was taken one night to a restaurant on a pontoon, floating in a mangrove swamp that is part of the Indus Delta. There was a full moon, and it shone on a flotilla of plastic rubbish, line astern, in the middle of the river. White in the moonlight on the ink-black water, it had neither beginning nor end. It floated slowly past us in complete silence, seemingly full of ominous purpose, grotesque but strangely beautiful, heading towards the Indian Ocean.

Acute mountain sickness can occur if you ascend above 2,500 metres. It is not clear why some people are more prone to it than others, or why the same person might develop it on one occasion but not on another. It is known, however, that the people of Tibet have different DNA from lowlanders, which enables them to live at high altitudes. Their DNA contains DNA from Denisovans – one of the early hominins that interbred with *Homo sapiens*, who now constitute humanity. Their haemoglobin binds oxygen more effectively than the haemoglobin of people who live at lower altitudes. Acute mountain sickness can be fatal if it results in cerebral or pulmonary oedema – waterlogging of the lungs and brain. Every year climbers die from it in the Himalayas. When working in Kathmandu I have seen brain scans of trekkers or mountaineers who had suffered from it, showing multiple small haemorrhages throughout the brain. In its milder forms it causes breathlessness, especially at night, with your breathing becoming characteristically periodic. You feel your breathing gradually speeding up until suddenly you stop breathing – apnoea in medical jargon – and then you take a deep, shuddering gasp. This makes sleeping very difficult. You start drifting off to sleep while your breathing slowly speeds up and you are then pulled violently awake by the apnoea, feeling that you are suffocating, gasping for air.

The explanation for this, in principle, is quite simple, although the details of the neural mechanisms involved are very complex and not fully understood. We inhale oxygen, and exhale carbon dioxide. The brain continuously monitors the levels of oxygen and carbon dioxide in the blood, keeping both at the correct level. At high altitude we have to breathe more quickly, as there is less oxygen in the air, but this lowers the levels of carbon dioxide in the blood, and the brain's balancing act falters. Our breathing then cycles between speeding up and then suddenly stopping. When it stops you wake from your superficial sleep with a shuddering gasp, desperately trying to get more oxygen into your lungs.

The teahouse we were sleeping in was very simple, without electricity and only an oil-drum stove for heating. It was minus five Celsius outside and a gale was blowing. There was a loose sheet of corrugated iron on the roof which rattled like a machine gun in the wind. William and I had a small cubicle to ourselves but the plywood walls were so flimsy that I could hear the man in the next cubicle, his head only a few inches away from mine, breathing and moving restlessly in his sleeping bag.

All this meant that I drifted in and out of only the lightest sleep. I could feel my thoughts gradually slipping out of my control as I sank into sleep, but then after a few minutes I would be jerked awake as I took a great, gasping breath. This had the curious effect that I felt as though the inside of my head had become a pitch-dark room and I was still half-awake and watching a rapid, hallucinatory slide show of thoughts, faces and abstract images. What was so striking was that I was sufficiently awake – at least, so it seemed – to see that these fragmentary images were entirely random. There was no meaningful sequence to them whatsoever that I could discern.

In recent decades, a whole new science of sleep has arisen. It depends on the technology of electroencephalography (EEG) – recording the electrical activity of the brain's surface through the skull and scalp – and on functional scanning (fMRI and PET CT). Until this new technology arrived, sleep was thought to be a time of rest and minimal activity in the brain.

In 1953 two researchers discovered – almost by accident – that there are periods when we sleep during which we rapidly move our eyes beneath our closed eyelids. Subsequent work has shown that sleep is strictly choreographed into REM (rapid eye movement) and NREM (non-rapid eye movement) sleep, with characteristic EEG patterns. There are typically five sleep cycles of approximately ninety minutes each. The earlier cycles are mainly NREM, but REM comes to predominate as the night wears on. There are usually four stages of NREM sleep, of increasing depth, where EEG shows that all the brain's waves of electrical activity are synchronised, often in a slow-wave form. With REM sleep, the EEG is desynchronised, highly active and irregular and indistinguishable from the waking state. If people are woken from it, they usually report coherent dreams – stories – whereas if woken from NREM sleep they report disjointed thoughts.

The content of dreams might seem random, but there is nothing random about the way we sleep. All mammals and birds have REM sleep. It has been conserved by evolution, as the saying goes, and must therefore be important. Sleep deprivation will prove fatal if sufficiently prolonged. Blind people do not dream in visual images yet demonstrate rapid eye movements. Predator animals sleep much longer in REM sleep than their prey. Dolphins sleep with one half of their brain at a time, long-distance migratory frigate birds do this as well. If people are deprived of REM sleep, they will enter into it increasingly quickly, as if they hunger for it. And deprivation of REM sleep reduces resistance to infection, showing that sleep matters for not only the brain. Babies and children spend much more time in REM sleep than adults.

It seems that the recently discovered 'glymphatic system' – the brain's lymphatic system which is thought to carry out cleaning and rubbish disposal – is especially active during sleep. Clearance of the amyloid protein which accumulates in Alzheimer's disease – its exact role in the development of the disease is controversial – can occur at night. Disturbed sleep is a well-recognised feature of Alzheimer's, although whether as cause or effect is not known . The list of such curious and fascinating facts is almost endless. There is much learned dispute about what all this means, and it is not even universally accepted that REM sleep is synonymous with dreaming. What is clear, however, is that sleep, and probably dreaming, is involved in learning and also in unlearning. Recent research implicates NREM sleep in decluttering and consolidating memory and REM sleep in reordering it, perhaps in new and creative ways. But the most interesting question to me remains unanswered – do my dreams mean anything?

Each time my breathing pulled me out of the lighter stages of NREM sleep, I heard the Himalayan gale blowing outside and the rattling roof, and I would briefly wonder about the strange slide show I had been observing in my own head, before the show started again and my feeling of being my own observer gradually started to unravel as I descended into the deeper stages of NREM. Was this a deep and meaningful view into the inner workings of my brain, or more like excavating middens – the ancient rubbish pits – that are such an important part of archaeology? That all I was seeing were the offcuts of my brain editing memories, or simply my brain passing the time, freewheeling, spinning like a gyroscope, which will fall over if it stops moving?

Opinions differ as to whether the slide show of light NREM sleep qualifies as dreaming or not. In the dreams of REM sleep we seem to be telling ourselves stories – sometimes as a participant, sometimes as an observer.

There is an intense sense of meaning and there appears to be a plot, as the dream unfolds over time. But what is also so striking is that other people's dreams are, on the whole, quite remarkably boring, although occasionally of curiosity value for their bizarre and random details. There is rarely any discernible narrative – the feeling that there is a plot is a mirage experienced only by the dreamer. It is as though we have a compulsion to make sense of things, turning them into a story, to see cause and effect, even when there isn't any. Functional brain scanning shows that the part of the brain associated with rational thought and analysis – the dorsolateral prefrontal cortex – is relatively quiescent during REM sleep, whereas the areas associated with vision, memory and emotion are highly active. A fact that makes sense, but that tells us nothing about whether the content of the dreams has any significance.

Freud did not discover the unconscious – the idea that we are not fully aware of why we act in the way that we do was not exactly new. He proposed an essentially hydraulic model of the mind: infantile erotic and aggressive drives from the id are repressed by the ego and superego. This build-up of pressure is relieved by dreaming, where the unacceptable wishes are transformed into an acceptable format. In other words, the mind scrambles the deep and unacceptable desires (sex with your mother, etc.) and the perceptive psychoanalyst, using the technique of free association, unscrambles the dreams. The content of dreams, in short, means something, although in coded form. Freud produced no evidence to justify his theory, so we don't need to provide any evidence to disagree with it. There are very obvious parallels with shamans and oracles, using their occult knowledge to decode dreams, often to predict the future – just as psychoanalysis developed many of the features of a religious cult.

But we have all experienced dreams that seemed pregnant with meaning, sometimes with horror, sometimes with love, where the relevance to our waking life seems quite obvious, albeit usually in a scrambled and bizarre form. I can still remember some of these 'numinous' dreams years later. The scientific literature abounds in studies of dreaming in depression, after PTSD, or in schizophrenia, and it is impossible to escape the feeling that dreams mean something, and are not just an epiphenomenon, like the ticking of a clock. There are many stories of famous revelations – Kekulé and the benzene ring, Dmitri Mendeleev and the periodic table of the elements, Paul McCartney and 'Yesterday' – that arose from dreams. But these revelations also came after much preceding conscious effort, and it is not clear if they arose in REM sleep or daydreaming. And all this brings us back to the metaphor problem – how are we to describe the relationship of the unconscious to the conscious? They are not separate entities, but parts of the same phenomenon. The difficulty we have in finding the words for this is not unlike how difficult – indeed impossible – we find it to understand the duality of light and matter as both waves and particles.

I lay in the dark at Gosainkunda, pulled in and out of

sleep by my brain's attempts to cope with hypoxia and troubled by my irritable bladder, although happily unaware of the cancer that was already invading it, longing to fall asleep. At times like this your consciousness becomes a wretched burden. My conscious self was an ignorant little boat floating on a great unconscious ocean, and I longed for it to sink under the waves, bringing sleep.

The misery of insomnia reminds me of what theologians call the problem of pain. If God is in charge, and is good, why does he permit so much suffering in the world? The theologians have never come up with a very convincing answer, other than to suggest that all evil is our own fault as we have something called free will, or simply that He works in a Very Mysterious Way. An entire discipline – theodicy – is devoted to creating a smokescreen to obfuscate this simple paradox. The let-out is that all wrongs will be righted in the afterlife. I cannot think of any religion that does not fail, if you take away belief in some kind of afterlife. As for neuroscience, if we and our brains are billions of nerve cells and synapses, complex electrochemical switches, why is pain painful? Why do difficult decisions *feel* difficult? Why is hypoxia so unpleasant? Why did evolution evolve *feelings*? Why can't I sleep, for God's sake?

The answer to this is very probably that when we – and probably most animals – make decisions, we use our feelings to guide us, just as much as reasoning. The classical distinction between reason and emotion is mistaken. They work together, not in conflict. There are a few people who have lost their amygdala to a rare degenerative disease called Urbach–Wiethe disease. The amygdalae are crucial to experiencing emotion, especially fear. Such people can reason logically but are hopeless at making decisions.

Once William and I had emerged in the morning from our sleeping bags, stiff, cold and under-slept, we stood looking down on the lake at Gosainkunda, grim and forbidding in the freezing dawn. We set off for the next high pass with our Sherpa guides. They said that they had also had a poor night's sleep. They are so strong and fit compared to me – although what with their Denisovan DNA there are good physiological reasons for this – that I was rather pleased to hear them complain. As we climbed higher, out of the cold and dark valley, we reached the sunlight and a staggering, celestial view behind us of the Himalayas. We could see for over a hundred miles – from Annapurna, past Manaslu and Ganesh, to Langtang Lirung and Helambu, with Tibet beyond. I do not know why the sight of such mountains resolves all anxieties and metaphysical problems, but it does. I strode on, amazed at how strong and well I felt, despite the awful night. I must have been deceiving myself, because William later told me that I was walking very slowly that day.

6

I suffer from a compulsion to make things – according to my mother, this showed itself at a very early age. I have no idea where it comes from. I have done most of the building work on my homes myself. In the early years of my first marriage, when I inflicted my building mania on my poor wife, there was some economic justification for it, as I was at first a medical student and then a junior doctor, working long hours on relatively low pay. There was no real need for it once I was a well-paid consultant, but I was unable to stop.

I once knocked down the brick wall between the two rooms at the back of our first home late at night, in order to build a new kitchen. When we fought each other twenty years later, as our marriage foundered, my wife cited this as an instance of my unreasonable behaviour – albeit one of the less egregious ones. Coming down in the morning, she said, to find the kitchen covered in brick dust, and the wall gone, was traumatic. Perhaps in retrospect she exaggerated, but I can see now that there was a kind of madness to my behaviour. When we sold the house ten years later, a surveyor for the prospective purchasers pointed out that the wall I had removed in fact supported the joists of the floor above and we were lucky that the ceiling had not come down. I fitted out the new kitchen myself, including a gas hob and the gas main under the floor. I had finally got the gas hob working at three in the morning and staggered out into the back garden in the dark, crying out (but not so loudly as to wake the neighbours) '*Fiat lux!*' I remember suffering from a severe migraine when at work later that morning. Shortly afterwards I tried to concoct my own furniture polish from beeswax, turpentine and carnauba wax over the new gas hob, without using a bain-marie, as I should have done. The mixture exploded, instantly filling the kitchen with toxic black smoke.

I moved into the house where I now live in the wake of the collapse of my first marriage in bitter acrimony twenty-one years ago. It is a standard South London Victorian semi-detached house, with the date 1887 in a circular niche under the front gable. Maps of that time show open fields with only a few streets laid out across them, and only a handful of houses. The houses are mostly 'two up, two down' – two rooms downstairs and two rooms upstairs, with an outrigger extension at the back for the kitchen and pantry and one more bedroom above. My house was built of London stock bricks, though the back of the house was unfortunately pebble-dashed at a later date. My brother researched the National Census for me from 1891 to 1911 and established that a printer called John Andrews lived here, with his wife and son, for at least twenty years. You possess your home so completely, it becomes so much a part of you, that it is hard to imagine that anybody else once felt the same about it. You feel a little threatened by the thought of their past presence, just as you resent the thought of strangers living in a house you once lived in yourself, especially if it was your childhood home. And then you wonder who will take your place, and to whom your home will be passed on. Just as I sometimes look at my hands and think about all the things I have done with them, and how one day they will be the cold white hands of a corpse, like the cadaveric hands I loved dissecting as a medical student.

The house is a five-minute bicycle ride from the hospital where I worked. I bought it from the recently bereaved widow of an Irish builder. I was told by my neighbours that in the months before his death, the old builder used to sit in the back garden, quietly watching the birds. The garden, when I acquired the property, was a wilderness, with ivy growing over the walls and festooning the trees, some of them long dead. There was a solitary rose beside a scrap of lawn, a camellia bush and an acanthus below it, and a crudely built cement-block shed backing onto the small local park. It was remarkably quiet, despite being in a busy London suburb.

I spent twenty years renovating and developing the house, doing most of the work myself and becoming more and more attached to it as a result. Over the front of the house there was a large loft space that I converted into an attic study. The sloping roof space over the rear extension – some four feet high – where the roof met the party wall, I used for storage, though access was only through a twenty-two-inch-square trapdoor.

The storage space became completely filled with

redundant possessions – it was less trouble to shove them into the loft than to sell them on eBay or take them to the local recycling centre. There was photographic darkroom equipment rendered obsolete by digital cameras, at least a dozen computers, each one replacing its predecessor as new models came out, yards and yards of cables, boxes of low-voltage transformers and lighting, old televisions, loudspeakers, LPs, hi-fi stuff, battered suitcases full of old clothes.

In a corner of the garage at the side of the house – another construction of mine with a leaking roof – next to stacks of wood that I have collected throughout my life, there was a six-foot-high tower of cardboard archival boxes. It was covered in cobwebs, and bird droppings from a pair of robins who had made their nest above it in an open box of workshop dust-extractor tubing. These archival boxes contained copies of all my private-practice medical notes for the seven years before I retired from private practice. For legal reasons I had had to keep them for seven years.

The end of the legal time limit for my medical records coincided with the onset of the lockdown. I found online a local company that shredded confidential documents. It seemed that shredding documents must have been essential work, as the business was open despite the lockdown. I drove round to the company's offices with the boxes. They contained thousands of pages, each patient neatly contained in a plastic envelope filed by Gail, my highly efficient secretary. I had to remove the paper from the envelopes to allow them to be shredded and knelt on the floor to do this. It took a long time. I was surprised at how many of the hundreds of patients I remembered – sometimes only the name, but often the diagnosis as well. On some of the notes Gail had written 'Died' with the date of death. Some of the patients I remembered very clearly indeed. Some of them, and their families, had suffered very greatly. Some I had got to know quite well, as they had slow-growing brain tumours that took many years to kill them, and I had seen them regularly as the disease slowly advanced.

It is a sombre experience to look at medical records outside of the clinical context of a hospital. They become statements of our vulnerability and mortality, which we cannot escape. Before cancer struck, I had various relatively minor medical problems – a broken leg, retinal detachments, a kidney stone. They were all treated very successfully by my colleagues, but whenever I saw my name in black and white, in copies of the correspondence sent to my GP, I felt a shiver of fear, even though the letters only recorded the fact that my treatment had been successful. I thought of the thousands of lives and conversations in my private-practice correspondence, about to be shredded. Patients rarely dare to tell doctors what they think of them and their behaviour, which makes it difficult for doctors to learn how to talk well to them. I can remember only one patient criticising me to my face, forty years ago, for the way that I had talked to her. I was doing an obligatory year of general surgery, before training in neurosurgery. Mrs Black, she was called – I remember her name because her courage in criticising me to my face was so unusual. She told me that I had told her too bluntly that she had breast cancer. She may well have been right. The lack of comment and criticism from most patients makes it all too easy for doctors to become pleased with themselves and have a mistakenly high opinion of their communication skills. Patients praise and thank us when things have gone well, but usually hesitate to comment when we have failed.

Like most doctors I liked to think that I was kind and sympathetic, but it was only when I was diagnosed with cancer myself that I could see just how great is the distance that separates patients from doctors, and how little doctors understand about what their patients are going through. Besides, as the great architect Frank Lloyd Wright observed, doctors can bury their worst mistakes and forget them, whereas an architect can only advise his client to grow vines over the offending building he has built.

Having sorted out the letters from the plastic holders, I got up painfully off the floor on my creaking knees and shrugged. Well, I thought, it's all over. I'll never know what my patients really thought of me.

As for the attic, it is remarkable what you can get through a twenty-two-inch-square opening, but the prospect of crawling on all fours back into the dusty loft to extract it was not a happy one, especially as I am getting stiffer and stiffer with age and had sometimes got stuck in the trapdoor opening as I pushed more things in. But with the lockdown I could stand it no longer, especially the feeling of shame and disgust at my extravagance and wastefulness over so many years.

Going in and out of the trapdoor made me think of Tibetan pilgrims crawling around Mount Kailash, gaining virtue through physical suffering. Or of Everyman in the eponymous mystery play (published in 1501), preparing himself for death, who finds himself alone, deserted by friends and possessions, as a ledger is drawn up balancing his sins against his good deeds. My sins, of course, are modern sins – sins against the environment – the excessive accumulation of material possessions, which then end up in landfill sites. Over a period of weeks, I emptied the loft out with the help of my son, who was living with me during the lockdown. We moved everything downstairs and filled up two rooms with all this stuff, and I eventually found a company that took it away.

I am not a hoarder – it's not that I cannot bear to part company with any of my possessions. Instead, I was lazy and greedy – constantly buying upgrades and new models and then too lazy to recycle things when I no longer needed them or get them mended if they broke down

Among the piles of clutter in my attic I found boxes of letters and juvenilia from more than fifty years ago. I found it hard to believe that I was so old. And many more boxes containing the diary that I have kept for over fifty years. What to do with it all? Part of me wants to shred the lot – to be free of all those pages, and to free my children from having to decide what to do with my diary after my death. I am quite certain that it is of no historical interest. But another part of me worries that there might be hidden treasures to be found, and when I do occasionally sit down and read a few fragments, I am fascinated by how much of my past I have forgotten, as well as how boring is much of what I have written. So I tell myself that one day I will plough through it all and keep a few fragments that may be of interest to my descendants. And yet to destroy most of it feels like a sort of suicide and I can't quite face it. I may well now leave it too late.

7

On my most recent visit to Ukraine, just before the pandemic struck, my colleague Andrij asked me to see one of his patients – I'm not quite sure why, as the man had recovered well from relatively minor surgery. He had suffered a gunshot wound to his head when serving at the Donbas front in the war against Russia and the separatists. He was a regular, professional soldier and a sniper. Evacuated from the front line, he eventually ended up under Andrij's care in Lviv. Andrij repaired the injury to his skull – he had been lucky that his brain had not been involved. Andrij told me that he had at one point summoned up the courage to ask the sniper if it was difficult to kill people. With a perfectly straight face the sniper replied that indeed it was, because his targets wouldn't keep still and moved around too much.

When I met the sniper, he was on leave from his injury but shortly to return to the front. Andrij acted as an interpreter. The sniper was shorter than me, with thinning blond hair and a round, rather boyish, innocent face, very clear blue eyes and a placid expression. There was a neat semicircular surgical scar two inches above his right ear. I think he was a little nervous and puzzled about meeting me. There had been some kind of crisis in his life, he said, but he did not elaborate. He had joined the army to sort himself out. He was not one of the nationalist volunteers who were playing an important part in the fighting against the Russians and separatists, but a full-time, proper soldier, as he put it. I remembered how forty-six years earlier I had become a doctor as a result of a crisis in my own life, and not from any deep sense of vocation, but I did not ask him to tell me more about his crisis.

'We always work in pairs,' he said, 'and we are very highly trained.'

'Is post-traumatic stress disorder a problem?'

'I don't think it is a problem for Ukrainians,' he replied, but I didn't believe him.

'Do you aim for the head?' I asked.

'Not necessarily. It depends on what you want to do. Sometimes you want to injure the man so that he lies in front of his comrades, crying out to them, but you stop them reaching him.'

'A very dirty war,' I said.

'Lasers are being used as well to blind soldiers.'

'I thought they had been banned,' I said.

'One of my mates was blinded in one eye when they targeted a laser on him. They are using landmines and white phosphorus as well. All banned,' he added, with a slight shrug.

'Do you hate the Russians?' I asked.

'No, no,' he replied with what seemed to be very genuine amusement. 'They're good guys, just like the rest of us.'

There was a pause in the conversation while I contemplated this.

'On a good day, how many people do you kill?' I eventually asked.

'We don't keep a tally,' he said quickly. 'Maybe the volunteers do, and make a notch on their rifle butt,' he added, in a slightly disapproving tone of voice.

I went on questioning him, trying to find some shred of feeling about his work, but he always responded that snipers were highly trained professionals – that there was no emotion involved in the work whatsoever.

'What do you think will happen with the war?' I asked.

'Soldiers are being killed every day. I see no end in sight.'

Had he really suppressed all his feelings so completely? Was he really so completely detached and indifferent to the people he killed? Or will he wake up at night in future years, full of dread and anxiety? Or simply find himself suddenly remembering his victims, in the way that in retirement I remember the patients I failed, often from decades ago?

On my visits to Ukraine, I would often see patients with large, difficult brain tumours called acoustic neuromas. These are benign tumours that in some patients can prove fatal if not treated. The problem is that surgery carries a high risk of causing damage to the nerve that controls the muscles to the face. If this happens the patient is left disfigured, with facial paralysis on the side of the tumour.

It might seem strange that I am so full of doubt in

retrospect about my abilities as a surgeon, but surgeons have to be judged by their failures – by their complication rate – not by their successes. It's surprisingly difficult to do this, as each patient, and each operation, is different and it is inevitable that some patients will do badly, however good the surgeon. And, on the whole, the better the surgeon, the more difficult the cases he or she will take on, so that the complication rate goes up. It's easy to hide and deny complications and mistakes – both from patients, from your professional colleagues and from yourself. I certainly did this at times, though there was at least one occasion when I could have done but did not.

I was assisting one of my trainees with an operation on a man's neck for a trapped nerve. The operation proceeded uneventfully but afterwards, as I was walking down the corridor outside the operating theatres, something was troubling me. And I suddenly realised, with a terrible sinking feeling, that we had operated on the wrong side of his neck. The operation is done through a midline incision, and it is usually impossible to tell from a scan after surgery which side has been operated on. The operation does not always relieve the arm pain for which it has been done, so it would have been easy enough to lie and recommend further surgery at a later date when his symptoms were no better. I know of one very eminent neurosurgeon who had done this, though with the much more major operation of a disc prolapse in the thoracic spine, when he had operated on the wrong level.

Full of dread, I went up next day to the side room where my patient was lying in bed. This was in the days of the old hospital in Wimbledon, which was surrounded by neglected gardens. It was spring, and his bed looked out over a green slope at the back of the hospital. It was covered with daffodils which I had planted myself the previous autumn.

'Mr Q,' I said, 'I'm afraid I have some rather bad news for you'.

'What's that Mr Marsh?' he replied.

'I'm afraid I've operated on the wrong side of your neck.'

There was a long silence as he took this in.

'Well, I quite understand Mr Marsh,' he then said. 'I put in fitted kitchens for a living. I once put one in back to front. It's easily done. Just promise me you'll do the right side as soon as possible.'

This was very many years ago, and long before the introduction of checklists (which would not be entirely certain to prevent the mistake occurring with the midline approach of this kind of operation). If this happened now, I would probably be dismissed. The pressure on me to lie would be all the greater.

Surgeons all learn to take the medical journals, full of amazing results that we could never replicate ourselves, with a pinch of salt. And yet there is always an element of sour grapes as well, driven by the fear – like Snow White's evil stepmother looking in the mirror – that perhaps there are other surgeons who are better than you are.

I once gave a lecture to the Acoustic Neuroma

Association – an organisation for patients with the tumour. Facing fifty people, many of them with partially paralysed faces, several of them my own patients, was one of the more unnerving experiences of my life. I felt a complete fraud, and the world's worst acoustic surgeon. At the end of the talk, one member of the audience – a young woman with red hair who before surgery had been an actress – came to talk to me. She had relatively severe paralysis of one side of her face.

'Your operation left me like this,' she said with a lop-sided grimace. 'But I could see that you were so upset when you saw me after the operation, that I forgave you.'

It had taken me many years to learn how to operate on these tumours and I was always worried that my results – in terms of facial paralysis – were not as good as the results published by other surgeons. But the surgeons publishing these results – mainly in the US and other European countries – were operating on far larger numbers of patients than I would ever achieve, and success in surgery is all about practice and experience. In the UK, surgeons have catchment areas more or less fixed by the NHS, and have difficulties developing the large, specialised practices you can find in some other countries. So you have to learn as best you can, and this involves both deception and self-deception.

All surgeons go through a difficult time at the beginning of their careers when they must pretend to their patients that they are more experienced and competent than they really are. In fact, this starts as soon as you become a doctor. There is nothing more frightening for a patient than a frightened doctor, and as a young doctor you are often frightened. So you have to hide your feelings from your patients. I don't remember anybody ever telling me this – you just learn it instinctively.

One of the great figures in evolutionary theory is the American biologist Robert Trivers. He has written a remarkable book entitled *Deceit and Self-Deception*. Deceit, he tells us, is universal throughout nature, as most living organisms prey upon other organisms lower down the food chain, and in turn are preyed upon by those above them. Even apex predators like ourselves or lions are food for bacteria. Deception is vital to survival, for both prey and predators. But what is remarkable about humans, Trivers argues, is our capacity for self-deception. As an evolutionary theorist, he explains this on the grounds that if we deceive ourselves when we are being dishonest, we are less likely to reveal our dishonesty through unconscious 'tells' and our body language.

When you are at the beginning of your surgical career you have to inflate your self-confidence, to deceive yourself, to enable you to cut into a fellow human being's body. And if you don't take on the difficult cases, how will you ever improve your skills? It is hard to admit to yourself that you have colleagues who are more experienced than you are, and to whom perhaps you should refer a patient with an especially challenging problem. Instead, we tend to deceive ourselves that we are more experienced than we really are. Self-deception, I like to joke, is an important clinical skill and perhaps does not matter too much when you are a junior in training supervised by more senior colleagues. But problems can arise when you become a senior surgeon yourself, and your work and decisions are much less likely to be observed. To find a balance between self-confidence and knowing when to ask for help, is one of the many tightropes on which all surgeons are precariously balanced.

Shortly after becoming a consultant, I saw a private patient – a young lawyer – with a small acoustic neuroma. This was over thirty years ago, when opinions varied as to whether small tumours need to be treated or not. (Opinions still do vary, to some extent, but there is now also the option of focused radiation treatment instead of surgery.) The patient asked me how many operations of this sort I had done. I remember becoming very defensive and embarrassed, as I had done scarcely any like his. This is the question, of course, that all patients should ask their prospective surgeon, but they very rarely do. I don't remember how I replied, but I know that at one point the patient angrily suggested I didn't want to operate on him because he was, in fact, a lawyer specialising in medical negligence. I ended up giving him the name of a surgeon elsewhere who was much more experienced at operating on these tumours than I was. It was obviously the right thing to do, but it was not as easy as it sounds, as it involves admitting your limitations both to yourself and to your patient.

It was because of cases like these that I had eventually fallen out with my long-time colleague in Ukraine. He had badgered me endlessly to teach him how to operate on large acoustic tumours, assuring me that there were no surgeons in Ukraine capable of treating them properly. A little reluctantly, I helped him with a few cases but eventually decided I should not – especially when I discovered that he was hiding some bad results from me, where patients had died or come to serious harm after surgery. The Soviet reflex of burying bad news was just too deeply ingrained in him. Besides, Ukrainian medicine was steadily improving, and other neurosurgeons were becoming increasingly skilled, and I no longer believed that my colleague and I could make a unique contribution to the treatment of these patients. So, after twenty years working together, we fell abruptly apart. A further problem had been that he would only share all the teaching and experience I gave him with his son. This was another part of the Soviet legacy – surgery in the Soviet Union had been profoundly nepotistic. If the state is against you, who can you trust other than your family and close friends? But this was very much against the culture in which I had trained. I continued to visit Ukraine, however, working with some younger doctors on simpler cases. Unlike my first colleague, they had not grown up in the Soviet era.

Olena, a young doctor, came to see me. The outpatient consultation took place in the bar of the hotel in central Kyiv where I was staying. I held the film of her brain scan up against the window. Through the scan I could see a view of pretty snow-covered streets outside. The hotel was in Podil, the only part of Kyiv surviving from the tsarist era. The family home of the writer Mikhail Bulgakov, the author of *The Master and Margarita*, was just round the corner in the next street. The young doctor had just given birth to a child, but during the latter part of the pregnancy had become a little unsteady on her feet and the brain scan showed a truly enormous acoustic neuroma. Its size meant that surgery had a very high risk of leaving her with life-changing, disfiguring facial paralysis. Apart from being deaf in one ear – the commonest problem with these tumours – she was remarkably well. The unsteadiness had improved after childbirth. Without treatment, however, the tumour was nevertheless a slow death sentence. She was determined that I should operate, and her family were raising the money needed for private treatment abroad. Treatment in Ukraine was possible but almost certainly more dangerous because of poor post-operative care. She asked what the cost of overseas treatment would be.

'In the UK, if all goes well, maybe $50,000. In Germany perhaps twice as much, and in the USA at least five times as much – probably more,' I said. 'And if you have problems after the operation, it could be many times more than that.'

'I want you to operate,' she said.

'I'm not necessarily the best,' I replied.

'But I trust you,' she said.

'I'll have to think about it,' I said.

It was not an easy decision. In the recent past, back in

London, I did not have to choose whether I, or somebody else, should operate. I had become the specialist for tumours like this and I would operate. But now, at the end of my career, there was a choice, just as there had been with the young lawyer at the beginning of my career. One of my younger colleagues in my hospital had taken over my acoustic neuroma practice on my retirement. He was a very different surgeon from me – quiet, thoughtful, immensely well-informed but modest, obsessional and infinitely patient. He was one of the few colleagues of whom I was a little in awe.

The case was potentially dreadful – death or disfigurement after surgery were real possibilities. If I asked my colleague to take on the case, was I being a coward? If I did the operation myself, would it be out of vanity and from a refusal to admit that my career as a surgeon had ended? I was certain that I could still operate well enough, but there was also my fear that my colleague might get a better result than I could.

Surgeons talk of 'sleepless nights' as a sort of code for the occasionally intense stress of their work. I had gone through a brief phase at the beginning of my consultant career when I would find it hard getting to sleep the night before a difficult operation. I remember sometimes drinking vodka with sleeping pills in the middle of the night – but it had quickly passed and from then on, I had always slept well, whatever awaited me the next day. But the night after that outpatient consultation in the hotel bar, I could not sleep.

I got up early in the morning. It was February, but the

winter was already fading. In recent years the winters have been becoming shorter and shorter with climate change. 'Climate change?' my Ukrainian friends say with a laugh. 'We have more immediate problems to worry about.'

The hotel was near the banks of the great River Dnieper, which runs through the centre of Kyiv. I went for a short pilgrimage to a little chapel nearby. It had been built shortly after the collapse of the Soviet Union, when churches and chapels began springing up all over Ukraine like mushrooms, after seventy years of suppression. This chapel lay on my route to work, and I had been driven past it countless times over the years. It was built on a small promontory of piles over the river, with a glittering, tinny gold roof, and was in pretty contrast to the bleak concrete motorway that separated it from the dilapidated pre-revolutionary buildings of the riverfront. I had often wondered what it might be like inside. As I walked towards it, I thought of how with my colleague I used to be very proud of my work in Ukraine and of how we had achieved great things. I now looked back on our years together in a different light. I realised that my achievements had ultimately been much more modest than I had thought at the time. I had chosen to see qualities in him that he did not really have and misinterpreted much of what I saw in Ukraine. I was blinded by my own vain wish to see myself in a heroic role. I now know that I had been wrong – neither of us were heroes. I don't regret the time I spent working there over so many years, but I look back with complicated feelings – a deep sense of failure, mixed with occasional triumph, muddled up with my love for the country and my many friends there.

I walked under the motorway that runs alongside the river, through a grim concrete subway, beneath roaring morning traffic. The pavements had turned into puddles of ice and rainwater. Everywhere there was the sound of melting snow dripping onto zinc roofs. The weather was abysmal – rain and fog, slush and dirty snow piled at the roadsides. The Dnieper was lost in mist, with a few broken floating ice floes, grubby white, cracked with black water. Three days ago, there had been little Brueghel-like figures fishing through holes in the immaculate, snow-covered ice, the entire river frozen over. There was a bent old lady in a headscarf in the dark chapel, standing in the shadows. The gilded iconostasis was only dimly visible. I bought a tall, thin candle for twenty *hrivnas*, and the old lady led me to a brass candle stand beside the iconostasis. Muttering in Russian, she lit two candles already on the stand. I lit mine and placed it upright, even though I have no religious faith whatsoever, and then walked back to the hotel.

I saw Olena and her husband later that day, once again in the hotel bar. I told them that I thought my colleague should do the operation. They were a little reluctant to accept this.

'Look,' I said, 'only one thing is certain – he can certainly get as good a result as me, and quite possibly a little better.'

Olena came to London a few weeks later, once the

visas had been arranged and £40,000 paid up front to the hospital. Olena had raised this by begging from family and friends. I put them up in my home to reduce the cost of their visit. The operation – carried out by my younger colleague and his ENT partner – took twenty hours. Twenty hours! They finished at six o'clock next morning. Olena emerged intact, her face spared, and was back in Ukraine after a month. It was a truly spectacular result. I was very proud to have such colleagues, but also a little sad.

Olena's operation – or rather the fact that I chose not to do it – marked the end of my career as an operating surgeon. Surgeons have a phrase for this – they call it hanging up your gloves.

8

I was frightened at the onset of Covid-19 that I might die, but what was absurd was that my main concern about dying was that I would leave unfinished the dolls' house I was building for my granddaughters. Thirty-five years ago, I made a dolls' house for my elder daughter Sarah. I was a junior doctor at the time, working a 'one in one' shift – in other words, I was on call continuously twenty-four hours a day, seven days a week. This was, in fact, illegal but fudged over by the hospital that employed me. It reflected the fact that the senior consultants were utterly unable to co-operate and would not allow their teams to look after each other's patients. I was perversely proud that this was probably the last such job in the country, and it appealed to my feeling of self-importance. I was able to live at home and the work was not that onerous at night – but it meant that I was severely restricted in my movements, having to be able to drop everything and go into the hospital quickly at any time. This was before mobile phones, and I had to carry an air call bleep. If it went off when I was not at home, I would have to find a telephone somewhere. I once had to go into a hamburger restaurant and beg the use of a phone. I gave my junior instructions as to how to insert a drain into a patient's brain, while the restaurant staff listened, obviously fascinated. I was rather annoyed when they demanded fifty pence for the cost of the call.

As far as I can remember, I made the dolls' house largely in the evenings and at night. My workshop was in a small cellar under the back of the house. There was a sealed-off lightwell with a pair of redundant windows in one corner, still with the criss-crossed strips across the glass to protect against bomb blast, dating back to the Second World War. The cellar was less than six feet high and I had dug out a shallow pit in front of my workbench so that I could stand upright. And there I made – more accurately, partially made – a monstrosity of a dolls' house, as inflated as my junior surgeon's ego. In retrospect, it was absurdly impractical – three storeys high and two rooms deep and four foot wide. And as with so many things I made over the years, I never finished it, and much of it was poorly and hurriedly made. Getting it up to Sarah's bedroom was a considerable problem – I remember irritably getting my first wife to help me, struggling up the stairs, pieces of the dolls' house breaking off as we dragged it up.

But despite its unfinished state, Sarah enjoyed playing with it, and made many dolls to live out their lives within its rooms. In time, of course, she outgrew it and the absurd construction followed me into divorce at the bottom of Wimbledon Hill. It sat, dusty and unloved, a sanctuary for spiders, in a corner of a back room for twenty years. I wanted to throw it away, but Sarah would not hear of it. Eventually I took a saw to it and amputated one-third. I told Sarah I would rebuild it for her two daughters but doubted if I ever would. And if it hadn't been for Covid-19, I doubt if I ever would have.

The problem was that I had made one part of the wretched thing quite well – the teak staircase, with two flights of stairs with sixty bannisters, two and a half inches long and a quarter of an inch square, turned on my old Myford lathe. It looked impressive but was useless for playing with dolls and furniture. The sixty bannisters must have taken hours of close-up work on my lathe to make – I was quite impressed when I looked at it twenty years later. I was surprised that my past self was capable of such work. Perhaps I was not quite so inept as I now choose to think. And I wonder whether my contempt for my past self is really a form of jealousy, driven by my awareness of how both my mind and body are deteriorating with age. Or that now, having retired, I have only myself to compete with.

The rebuilt dolls' house was a great success with my granddaughters Iris and Rosalind, and so I started work on an entirely new dolls' house for my youngest granddaughter Lizzie. The first dolls' house had been a version of the early eighteenth-century terrace house my family had lived in after we moved from Oxford when I was ten years old. I suppose I was trying to recreate my past in miniature, to return to a better time. The second dolls' house became a castellated, multistorey fantasy, with another grand staircase, and with all the external walls held in place with magnets so that they could be removed for easy access to the rooms inside. I found, at long last, a use for some of the wood stored in the garage – the burr elm made a fine marbled floor for the basement, and the bedroom floorboards are of ash and ebony. There is also a hammer-beam roof made of oak that took many days to make, with miniature finials that I turned on my lathe. The lockdown and retirement meant that I had plenty of time to do the work properly, but I still haven't been able to resist making it too large. My daughter Katharine assures me there will be room for it in their home, but I fear that as time passes, it will become more of a nuisance than an heirloom.

Some time ago I built a lean-to garage at the side of my house. As with too many of the roofs I have built, the garage roof leaks badly when it rains. The wallplate timbers are starting to rot and weeds are starting to grow through the roof – they look quite decorative. It all needs to be redone, but I constantly postpone this. The garage is now filled with wood. I have been collecting wood for much of my life. I like to look at it, stacked up in my garage, and think of all the possible things within it, which I have yet to make. I can tell you the history of each piece. Steamed Swiss cherry wood from a timber merchant in Limehouse forty years ago. Huge pieces of burr elm, with convoluted, mamillary surfaces, and dense wild grain. They sat for years in my garage until I realised that the wood could be used to make beautiful marble-like floors for the dolls' house. Elm wood is now very hard to find as the trees came close to extinction in Europe 50 years ago as a result of Dutch elm disease. I felt quite disconcerted a few years ago when I saw, in the Botanical Gardens in Christchurch, New Zealand, a great English elm tree – its distinctive shape brought back childhood memories of English hedgerows and family picnics in summer.

There are huge discs, like Parmesan cheeses, of spalted beechwood. Spalting is the elegant pattern of black lines that is produced by fungal infection in beechwood. It looks especially attractive in turned bowls. I had found the remains of a great, fallen beech tree in a valley in mid-Wales and with my 30-inch chainsaw had sawn off some large blocks. I subsequently cut them up into large discs on my bandsaw with a view to turning them into bowls. But I have never done this, and they have sat in my garage for years gathering dust and cobwebs and, alas, woodworm holes. When I look at them now, I think of the silent and remote valley, and the scent of damp, fallen leaves and the sound of the nearby stream – an almost primaeval scene, were it not for the smell of oil and petrol from the chainsaw. There is applewood from a farm in Kent and mulberry wood from the garden of the widow of the consultant neurosurgeon whom I replaced on his retirement. There is a large quantity of cedar of Lebanon timber from the gardens of the hospital in Wimbledon where I worked for many years. The tree had been killed when asphalt was laid down for a car park around it. Cedar wood has a beautiful scent which repels moths and is excellent for lining chests and drawers. They are grand trees, which, when mature, spread out like giant umbrellas, with the lower branches dying off. They used to cover much of the Middle East, but now are confined to a small fenced-off enclosure of a few dozen trees in Lebanon. When I visited the place – a pilgrimage of sorts – the trees made me think of sad animals, soon to be extinct, caged in a zoo. I have grown one cedar of Lebanon in a large terracotta planter – it seems a little cruel to confine like this such a mighty creature but after twenty years it seems happy enough with green shoots this spring. Pot-bound, it has adopted the umbrella shape of a fully grown tree and has become a giant bonsai.

There are many short billets of exotics – rare tropical hardwoods such as cocobolo or African blackwood – which I bought years ago to turn on my lathe but have yet to do. They are now largely unobtainable or prohibitively expensive as a result of the decimation of the rainforests. Nor, out of shame, would I buy any now, even if I could. I even have a few thick leaves of sandalwood veneer, which I bought from a luthier almost fifty years ago, who had found no use for them. I'm not sure where he, in turn, had found them. Trade in that wood has been strictly controlled for many years – the wood is mainly used in the fragrance industry. I gave some of it recently to a woodworking friend who makes small boxes. The veneer has lost much of its fabled scent, but I hope that when it is sawn and sanded the deeper layers of the wood will still have it.

I am constantly having new ideas of things to make with all this wood – but the fact of the matter is, whatever happens, I will not live long enough to use even a fraction of it. I would look at my hoarded wood with deep pleasure, but as old age and decline approach, this pleasure is starting to fade and instead is replaced by a feeling of futility, and even of doom – of the future suggested by my brain scan. Besides, anything I now make will outlive me, and I should only make things that deserve to survive in their own right. I no longer have the excuse of the craftsman – who sees all the faults, often invisible to others, in what he has made – that I will do better next time.

PART TWO
Therapeutic Catastrophising

9

I had been planning on seeing a medical colleague about my increasingly irritating prostatic symptoms – poor flow, and urgency and frequency of urination – but the lockdown put this on hold. Besides, the lockdown was such a strange and intense experience that I quite forgot my symptoms and another seven months passed before I arranged an appointment to see a colleague. To save time I decided to go privately, although I no longer had private medical insurance.

Having carefully washed my bottom, in anticipation of a rectal examination, I cycled into Harley Street, swigging a litre of mineral water as I went. I had been told to do this so that I could have my urine flow measured on arrival.

I had to report to a friendly nurse who made me drink many cups of water. She would put her head round the door every so often.

'Are you bursting yet?' she would ask.

When I eventually reached this point, I was directed to a urinal which carried out the necessary measurements and recorded my sad and struggling attempt to empty my bladder – a problem I had been living with for many months, perhaps even years. Once this was done, I was ushered up a grand carpeted staircase to the consulting room.

The double oak doors of the room were so tall and imposing that I hesitated to go in, finding it hard to believe they were simply for a medical consulting room, but this was Harley Street, and not the NHS. The room was huge and my colleague Ken, masked like myself for the pandemic, was sitting behind an enormous desk. It reminded me of stories of Mussolini, who had a gigantic desk in his office. His Cabinet ministers had to run at the double the long distance to his desk when they came to deliver their reports. But Ken is a very nice man and not at all like Mussolini. He had operated on me two years ago for a kidney stone – I had made careful enquiries as to whom I should consult. Being able to do this is probably the greatest benefit of being a doctor yourself. It is otherwise less clear that being a doctor is helpful when you are ill. In my case, it proved to be little short of disastrous.

We chatted for a while. The Covid crisis had been good for him, he said – his NHS hospital had come to understand that stones, as he put it, were important. Patients continued to need urgent treatment for kidney stones during the lockdown, unlike some other specialties. We discussed my symptoms – I found myself playing them down, or at least my endless preoccupation with them.

'I need to examine you,' he said a little apologetically. In the past I had always rather dreaded having a rectal examination – in practice it is unremarkable.

'Your prostate is a little firm,' he said, as I pulled my trousers up.

'I don't want a PSA,' I said. PSA stands for prostate specific antigen, and is an abbreviation with which many ageing men are deeply concerned. The test measures a protein in the blood that is secreted specifically by the prostate gland. The prostate steadily enlarges in most men throughout their life, and in one in seven men turns cancerous. In these cases the PSA will rise, although cancer is not the only cause of a raised PSA, and a slightly raised level in an older man can be perfectly normal. For many men the cancer is relatively harmless – they die with it rather than from it, with few ill effects. This can make it quite difficult to decide whether to treat the cancer in every case or not – as no treatment is without some risk. The cancerous gland can be completely removed with surgery, provided it has not spread beyond the gland's capsule, but the operation comes with the risk of impotence and incontinence and it can be hard to know when the risk of surgery is justified. But if the gland has spread beyond the prostate, it will probably kill the man – although this might take some years.

Ken managed to persuade me to have a PSA test. I couldn't very well deny that I had come to seek his advice.

'If it is cancer, I don't want any treatment,' I told him, 'unless it progresses.'

'I know where you're coming from, but it's no good putting your head in the sand,' he said.

Looking back, I am completely amazed at how wilfully blind I was – how I had been so frightened by my symptoms over the years, that I had refused to admit the need for a PSA, and had now probably left it too late.

'You know,' I said, as I was about to leave, 'when I was still in practice, all I ever wanted to do was operate all the time. It meant more to me than anything else, although I also loved caring for patients. But now that I have finished, I don't miss it at all – I'm not entirely sure why not. Do you like honey?'

He replied that he did, and that he had honey every morning for breakfast, so I pulled out the small pot of honey made by the bees I keep in my garden and gave it to him.

I had had intermittent prostatic symptoms for close on twenty-five years, which at first were almost certainly due to a common condition called chronic prostatitis. I was a little embarrassed by them, and did not seek professional help, and also as a doctor I suffered from the firm conviction that illness happened to patients and not to doctors such as myself. When we are medical students we enter a new world – a world of illness and death. We learn about all manner of frightening diseases, and how they usually start with trivial symptoms. Many students, in response to a few minor aches and pains, become convinced that they have developed a catastrophic illness. In order to survive, they have to believe that diseases only happen to patients and not to themselves. A few doctors remain hopeless hypochondriacs throughout their careers, but most of us carefully maintain a self-protective wall around ourselves, which separates us from our patients, and becomes deeply ingrained, sometimes with unfortunate results. Doctors with cancer are often said to present with advanced disease, having dismissed and rationalised away the early symptoms for far too long. I was well aware of this phenomenon, but this knowledge did not prevent me from falling victim to it myself. Prostatism affects most older men – in medical language, frequency and urgency of micturition, and poor flow. In medical school, students are taught a process called the diagnostic sieve. Any set of symptoms can be caused by different pathologies. There is an acronym for these – MIDNIT – metabolic, inflammation, degenerative, neoplastic, infection, trauma. And the neoplastic can be benign, or it can be malignant; cancer, in short. Inflammation of the prostate cannot be distinguished from cancer in its early stages. In theory I knew this, but for too many years I had indeed chosen to bury my head in the sand. I was a doctor – admittedly a retired one – and illness still only happened to patients, not to me.

So when the simple PSA blood test showed that I had a PSA of 127, I couldn't really believe it. Only 4 per cent of men with cancer of the prostate present with a PSA over 100 – most cases of cancer will be well below 20. Frantic, panic-struck googling told me that most men with a PSA of over 100 will be dead within a few years.

I was referred to a famous NHS cancer hospital, the Royal Marsden, in central London. It was six miles away from my home, and as I had read that cycling can put up your PSA from the pressure of the saddle on your bottom, I walked to the hospital. I hoped that this would show the first PSA reading was a mistake, and not a death sentence after all.

Forty years ago, when my son William was lying in hospital at the age of three months with a potentially fatal brain tumour, my first wife and I were so sick with anxiety that the outside world seemed to disappear – at least to become unreal and insubstantial, or perhaps it was that we ourselves had become ghosts. I envied people I saw around me, and what I imagined to be their happy, carefree lives. But on this occasion, at the age of seventy, with my own life under threat, much to my surprise, I found that I felt deeply sympathetic to all the people I passed by as I walked to the cancer hospital. I did not envy them in the least and hoped that they would have as good a life as I had had. I walked to the Thames along the River Wandle Nature Trail. There was a young woman beside a noisy weir, fishing.

'Any luck?' I asked her.

'No,' she replied. 'There are brown trout in the river now. But I wouldn't want to eat them,' she added.

We chatted for a while and I then continued to the railway station at Earlsfield and on to the Thames, where I could join the Thames Path, with its views over the river and the many new tall apartment buildings that now line it. I strode along, occasionally looking at my new and expensive brogue boots that I had especially polished for the six-mile walk. I remembered my brief time as a gynaecological junior doctor forty years earlier, and how the women attending the abortion clinic in the morning were dishevelled and downcast, whereas the women attending the infertility clinic, which was in the afternoon, were all immaculately turned out. It will all prove to be a mistake, I told myself. My PSA was so high because I bicycled in to have the test. There has been no research into seventy-year-old men with chronic prostatitis, who ride bicycles. Telling myself these comforting fairy stories, reassured by my beautifully polished boots, I crossed the ornamental Albert Bridge to reach the well-heeled streets of Chelsea and then the hospital.

The Marsden (which at work we called the Mars Bar) was built in 1851, and was the world's first dedicated cancer hospital, named after its eponymous founder. I had delivered the annual memorial lecture there myself two years earlier. I had been the hospital's neurosurgeon and in previous years had operated on many of their patients, who were sent to me in my own hospital – St George's in Tooting – if they had cancers in the brain or spine. So the place felt quite familiar. Besides, most of my adult life has been spent in hospitals.

There were the usual long, brightly lit but windowless corridors from which all hospitals suffer. And a few pathetic works of original 'Art' on the walls, along with many admonitory notices. At least everything was spotlessly clean. The outpatient waiting area was less subterranean than the one in my former hospital, and almost empty. The waiting area for the scanning department, to be fair, which I went to the following week, was rather wonderful (although with the usual drab works of original art and notices). Scanners weighing many tons had somehow been hoisted up into the attic space and the waiting area had a huge skylight looking out onto roofs and an ornamental cupola. Rain was falling when I went for the first of my various scans. Looking at the grey slates and lead of the roofs, shining with rain beneath heavy clouds, made waiting a pleasantly philosophical experience, despite the other poor patients waiting alongside me, some of whom looked ravaged by chemotherapy and disease.

On one of my journeys to Ukraine I had operated in a hospital in Odessa. It was a private hospital, which had formerly been a fruit-machine factory. Some years earlier the Ukrainian parliament had banned gambling machines and the factory's owners were left with a problem. They decided to turn the factory into a private hospital. It must be the most windowless hospital that I have ever seen, and still felt like a factory, but clean and white. Not even the patients' rooms had windows. But in one subterranean area an entire wall had been fitted with a floor-to-ceiling backlit photograph of the sun rising through trees. It completely transformed the dull room and inspired me to have a similar picture wall installed in the windowless outpatient waiting area in my own hospital. I managed to arrange this, with the agreement of the hospital management, using charitable money that I had been given by patients over the years. The photograph was by the landscape photographer Charlie Waite, many of whose photographs I had hung on the walls of the neurosurgical department a few years earlier. The backlit photograph replaced two gloomy abstract canvases, pastiches of the suicidal artist Mark Rothko, in purple and dark scarlet, which could surely only serve to reinforce the feelings of impotence and anxiety most of us feel as we sit in outpatient waiting rooms. Research in America by Roger Ulrich – the founder and doyen of the study of the impact of the hospital environment on patients – has shown that what we want to see, when ill and anxious in hospital, are pictures of landscapes, ideally with water and paths leading away into the sunlit distance. Or smiling faces.

My previous attempt to improve the hospital's appearance had got me into a certain amount of trouble as on that occasion I had not sought anybody's permission. I had gone in to the hospital with a hammer and nails one weekend and hung up many large framed copies of Charlie Waite's landscape photographs. The building was built under the private finance initiative (PFI), and so was owned by a private, highly profitable company that rented the building to the NHS. PFI was supposed to be a cheaper and more efficient way of building hospitals and schools than direct government funding, but as many warned at the time, the opposite was true. PFI was a con, little short of an economic crime, for which nobody has been held to account. Fortunately, I knew the manager of the building well, and had treated his son, so I was forgiven. The pictures are still there on the walls today, although I was told I should have sterilised the walls behind the pictures before hanging them. I have no idea whether my modest contribution to the hospital's corridors has made any difference to anybody, or is even noticed, but I enjoy seeing them there whenever I go into the hospital.

Much of what goes on in hospitals – the regimentation, the uniforms, the notices everywhere – is about emphasising the gap between staff and patients, and helping the staff overcome their natural empathy. It is not about helping patients. Hospitals always remind me of prisons. These are places where your clothes are taken away, you are given a number and you are put in a small, confined space. You must obey orders. And then you are subjected to a rectal examination – well, perhaps not always. As my anthropologist wife Kate – who has been in hospital more frequently than she would like – tells me, patients often ask each other exactly the same question as prisoners: 'What are you in for?'

I didn't really understand hospitals until I met Kate. As an anthropologist she saw very clearly things that I had never, to my shame, noticed. She pointed out to me that the last thing you get in hospital is peace, rest or quiet, and that being a patient is an essentially disempowering and humiliating experience. My meeting Kate also coincided with the old nineteenth-century hospital – Atkinson Morley's in Wimbledon, where I had worked for many years – being closed. My department was moved to a new building in a huge teaching hospital. The contrast was extreme.

The word 'hospital' derives from the Latin word *hospes* – meaning guest. In the early Middle Ages, 'spitals' were charitable refuges – places of hospitality – usually in monasteries, for travellers, the ailing and the poor. The administration of the holy Mass was just as important, if not more so, than what little medical treatment was available. The first dedicated hospital was built by the architect Filarete in Milan in the late fifteenth century. It had the same cruciform design as a church, so that the patients in their beds could all see religious services at the crossing. After death, the patients were placed in a crypt beneath the crossing, but the smell became intolerable, and a separate graveyard had to be established. Early hospitals must have been dominated by foul smells, despite the prevailing miasmatic theory of infection, dating back to Hippocrates, which held that infections were spread by foul air rather than by physical contact. (It was correct to the extent that respiratory infections – like Covid – can be spread in the air, but they don't smell and most infections are spread by contagion, by touch.)

It was not until Florence Nightingale in the nineteenth century that hospitals started to develop their characteristic design. She was a firm believer in the miasmatic theory and had a major and beneficial influence on hospital design in Britain, albeit for the wrong reasons. Atkinson Morley's Hospital (known to everybody as AMH) was a Nightingale Hospital, with tall ceilings, and high windows – the emphasis being on fresh air and light. It was built before hospitals became machines for healing, and when Wimbledon was still surrounded by fields and gardens. The hospital had had its own stables, and laundry and staff residences. This had slowly been eroded, and some of the land sold for development, but when I worked there it was still surrounded by fields and trees, some of them mighty cedars of Lebanon and oaks. It was only three storeys high and had a staff of fewer than 200. It was built on a human scale. I knew all the staff – not just the other doctors, but the nurses, the physios, the porters and cleaners. I think all of us, at all levels, had a real sense of belonging and personal responsibility for what happened to our patients. The hospital was famously efficient – I would get through three or four major surgical cases a day, something unimaginable now.

We are tribal animals – we are happiest in relatively small groups. When I walked into the old hospital – a bit dilapidated, it is true – I would know everybody I passed in the corridors. I felt at home. With the new building, when I entered, I recognised scarcely anybody. There was no sense of belonging. Everybody hurried past, trying to get to their own little oasis. I did my best to try to recreate some of the features of the old hospital in the new building. I equipped the registrars' on-call room with an Afghan rug, a bed and armchairs but after a while the management declared this a fire hazard and it was all removed and replaced with hard office furniture and no bed. (The armchairs were, in fact, labelled as fire-resistant, but never mind.) The Afghan rug – quite a good one – disappeared. The old hospital is now a gated estate of luxury executive homes and apartments.

The new building was built with wide balconies outside the wards. We all know, of course, that the first thing a patient wants to do in hospital if they see a balcony is to throw themselves off it, and nobody was allowed onto them. After years of struggle on my part with the management and raising large sums of charitable money with my colleagues, it was possible to make the balconies outside the two neurosurgical wards suicide-proof, so they could be turned into a healing garden, immediately accessible to the patients from their beds. Patients can – a brief reprieve – escape the prison and see the sky and green plants. There are elegant pencil cypresses – now fifteen feet tall – growing in large tubs, and creepers climbing up the hospital brickwork. There are plenty of chairs, sofas and even sunloungers among the many planters. The view is over the slate roofs of Tooting and is much more beautiful than you might expect. It is true that many hospitals now have healing gardens, but they are almost always a long way from the wards. Most patients do not want to walk down long corridors trailing urine bags and drip stands with them, so all too often there is little healing. Creating the neurosurgical garden has probably given more relief and happiness to more people than anything else I have done. It is now cared for by a charity run by a late patient's family, and there are plans to turn all the other balconies on the dull building into gardens. It will cost a lot of money but will be wonderful. When you approach the hospital, you will be met by a living, green wall.

It is difficult to show in a rigorous, scientific way that

we are hardwired by evolution to feel happiest when surrounded by green, living things. I have a friend who grew up in New York and has absolutely no interest in flowers or trees or landscape. But I would like to think that the love of Nature is innate in most of us. So why are most hospitals so often so horrible? Why is it that it is only when we are dying in hospices, that gardens and flowers and trees are allowed to re-enter our life, just as we are leaving it?

I have thought about this question for many years. There is no simple answer, but what is certain is that it is very difficult to introduce environmental enhancements after a building has been built. Especially when, as is so often the case, the hospital is built in a confined, urban space. You need to plan the enhancements from the beginning, and this rarely happens. There is a standard maxim in architecture that the secret of a successful building is an informed client. It is easy to blame architects for a poor building, but ultimately the quality of the building is determined by the people commissioning it. But who is the client for a hospital? The patients? The managers? The healthcare staff? The government? I have always been very struck by how so many hospital managers and doctors have no interest or understanding of the importance of good design. They see hospitals as little different from machines, and yet at least 75 per cent of the lifetime cost of a building is the cost of the staff working in it. If you build a fine building, the staff will work more efficiently, and take less time off work – it will save money in the long term. And patients will probably recover more quickly. But this kind of longer-term thinking has been sadly absent from the NHS. And beyond this, despite all the notices on the hospital walls declaring that patients are treated with dignity and respect, patients are still seen as an underclass, and trying to improve the quality of the hospital environment as a waste of money. And if patients really were treated with dignity and respect, there would be no need for all those notices.

At the Marsden, once I had been checked in by an unsmiling receptionist, I sat down beside a stand of pamphlets about living with a wide variety of cancers – prostate, rectal, breast, pancreatic. They had pictures on their covers of healthy-looking elderly people smiling manically. I wondered whether they were models or actual patients. A nurse eventually came, and I was weighed and measured. I noted that I was almost two inches shorter than when I was a young man, and much to my annoyance that my bathroom scales had been flatteringly underestimating my weight by five kilos. I was then told I needed to perform once again on a urine-flow device. I was put in a small side room and presented with many plastic cups of water, which I dutifully drank before being led out like a child to the specially equipped toilet.

I emerged a few minutes later holding the printed read-out which measured objectively my difficulties urinating. The nurse glanced at it briefly with a rather disapproving look. I got the distinct impression that I had not tried hard enough. I felt as though I was entering my second childhood already and that I was being potty-trained all over again.

I followed the disapproving nurse back to the side room. She had long, luxuriant dark hair down to her waist.

'I like your hair,' I said.

'I am growing it for charity,' she replied. 'To make wigs for the women having chemotherapy.'

I had not received a word of explanation about what was happening until, as she left the room, she told me that the doctor would be coming to see me.

After a while the oncologist arrived.

'Let me start by saying how sorry I am that we are meeting like this,' he said.

I suppose it was kindly meant but I found this rather a depressing start to our relationship, and it filled me with foreboding. He spoke to me for a few minutes and assured me that he would fast-track the various scans that were needed to establish whether my cancer was already widely spread or not.

'How probable is that, given my PSA?' I asked.

'70 per cent,' he replied, looking away from me.

I asked hopefully about the effect of bicycling on my PSA.

'You would have to bicycle a hundred miles on a very bumpy road to raise it by maybe one,' he said.

I struggled with being both a doctor and an anxious patient at the same time and found it very hard to ask him about my future – reluctant to hear bad news but hoping for hope.

'Please talk to me as a doctor,' I said to him. 'I used to have to tell my patients about their cancers and try to cheer them up at the same time.'

'That's not how we do things here,' he replied cryptically. In retrospect I realised I had given him conflicting messages – that I wanted to be told the truth but also given hope.

He was sitting perched on the edge of a chair, as though he was about to leave any minute, with a piece of paper on his knee, on which he jotted down a few notes. I found myself feeling awkward and tongue-tied. I inevitably blurted out to him the question that all of us ask oncologists when we first meet them: 'How long have I got?' – or rather a medicalised version of it. I asked him what the probabilities were that I would be alive in five years' time with a PSA of 130 as the only predictor. In fact, I already knew the answer: 30 per cent. But he did not tell me this.

'You needn't write your will for five years,' was his reply. The reality, of course, is that he could have no idea what would happen to *me*. I knew this, but still, childishly, hoped he would tell me that I would be fine. He could only quote probabilities, which he seemed reluctant to do. Patients want certainty, but doctors can only deal in uncertainty.

'Let's get to know a little about you,' he said. I said that I valued being physically fit and that I wrote.

'If you write one book a year, you will be able to write five more books,' he said with a laugh. Perhaps he was trying to reassure me, but I felt he underestimated the difficulty of writing.

'I read somewhere that hormone therapy can have cognitive effects,' I ventured.

'You may be a little less sharp,' he replied but did not elaborate. He may well have told me more about the possible side effects of treatment, but if he did, I was far too anxious to take them in. I had always known, as a doctor, that patients only hear a small part of what you tell them, especially at the first visit.

He mentioned something about my meeting 'the team' and then left.

Percentages are a problem for patients. Some of the oncologists I have worked with over the years told me that they would never give patients percentages. The problem, of course, is that the patient wants to know what will happen to him or her as a specific individual, and the doctor can only reply in terms of what would happen to a hundred patients with the same diagnosis. After a given number of years a certain percentage will still be alive, and the remaining percentage will be dead. There is no way of knowing into which group an individual patient will fall. Your doctor never knows how long you will live, not until the very end.

When I thought back on my years as a surgeon, often dealing with cancer, I realised that I, too, rarely talked in terms of percentages. Malignant gliomas – primary brain cancers – have a mortality of at least 50 per cent at one year, and only 5 per cent or so of patients are alive at five years, despite treatment with surgery and radiotherapy.

I told patients with these tumours that if they were

'unusually unlucky' they might be dead in six months, and if they were 'unusually lucky' that they might be alive in several years' time. I would explain that for most people the tumour would recur between these two extremes, and that further treatment might be possible, without admitting that further treatment usually achieved very little.

At the time I thought that this was quite a good way of dealing with the problem, and of finding a balance between hope and realism. In the days of Google and the Internet, I am not sure if this is still true.

Two months before my own diagnosis of cancer, a very good friend of mine – Phil Rogers, a wonderful potter – was diagnosed with a malignant glioma. He had become confused over the course of only a few days and his wife Ha Jeong, also a superb potter, had rung me in desperation, as she had been unable to get the local hospital to take the problem seriously. It was all made more difficult by Covid and the fact that my friend had no insight into his problems, as is usually the case with tumours affecting the frontal lobes of the brain. I told her what to tell their GP and a brain scan was subsequently obtained. To my dismay this showed one of the most lethal brain tumours – with the seemingly benign name of 'butterfly glioma'. (On scans, the shape of the tumour roughly resembles a butterfly.) These tumours start in the corpus callosum – the millions of nerve fibres connecting the two cerebral hemispheres. The tumours rapidly spread into both sides of the brain and cause progressive confusion and dementia within a matter of weeks. All patients with these tumours are dead within a few months, and treatment achieves nothing. High-dose steroids temporarily reverse the symptoms, but only for a few weeks.

I had always felt very strongly that I would not want any treatment if I had a brain tumour like this. Steroids for a few weeks, yes, but no radiotherapy and no surgical biopsy. And yet this is the standard treatment for such tumours, even though I don't know any neurosurgeons who think it achieves anything. I had little choice other than to talk to Phil and Ha Jeong. I told them to accept the advice they received from my colleagues. This was, inevitably, for biopsy and radiotherapy. As they lived in a remote corner of mid-Wales this involved many long journeys, and as I was soon to discover myself, albeit with just a one-hour bicycle ride, the journey to hospital for treatment is often worse than the actual treatment. But although I knew the treatment was a waste of what little time Phil had left, I felt unable to tell him that there was nothing to be done, to deprive him and his wife of any hope, even if hope for only a few months. I sat down with them and explained that the tumour would be fatal, and that the next few weeks would be the best time he had left, and he should make the most of it. I said that if he was lucky he might be well for many months.

He took this all in with remarkable equanimity. 'It is what it is,' he said. It was difficult to know whether this came from stoicism or frontal brain damage. He was a world-class potter, potting in the Leach–Hamada tradition, and there are pieces of his in museums all over the world. He made the most beautiful bottles and pots, and his barn where he stored the finished pieces was like an Aladdin's cave. Whenever I visited him, I would explore the cave, finding wonderful things. I had bought many of his pots over the years. Shortly before he fell ill, he had fired his wood-fired kiln. The results of wood firing are often unpredictable – the kiln has to be kept going for forty-eight hours and it is difficult to control the temperature. When a few days later the kiln has cooled down and is opened, there is great excitement and anxiety as to how well it has gone.

Phil opened the kiln a few days after I talked with him.

'The best ever,' he said. 'Rather ironical that this is probably the last one. But it will be a nest egg for my wife.'

It was indeed a superb firing. After his death Ha Jeong stored all the pots from this last firing in the barn. A few months later, and after the easing of the Covid lockdown, she organised a large party to celebrate his life. A few days later, thieves, who must have heard of his death, broke into the barn and stole the most valuable pieces. Ha Jeong had not had time to photograph or catalogue them, and so had no way of knowing exactly what had been stolen. The police were therefore unable to help. She later gave me a small vase from this last firing. A beautiful piece with a flecked grey glaze (Phil was particularly good at glazes), it sits in front of me as I write. I look at it and I like to think of his hands shaping it, as it spun on his potter's wheel.

When I now think of how the uncertainty about my own future, and the proximity of death, threw me into torment, careering wildly between hope and despair, I look back in wonder at how little I thought about the effect I had on my own patients after I had spoken to them. I did worry that if my tone of voice was too pessimistic the poor patient might spend what little time he or she had left feeling deeply depressed, simply waiting to die. So I tried to find a balance between telling them the truth and not depriving them of hope. After a patient died, I only occasionally heard back from the family, so I had little way of knowing whether the way I had spoken to them was appropriate or not. As I was discovering myself, false hope – denial by another name – is better than no hope at all, but it is always very difficult for the doctor to know how to balance hope against truth when talking to patients with diseases such as mine.

I must have misunderstood the oncologist about meeting the team, because when the nurse returned to say that I could go, I said that I thought I was going to meet the team. The nurse looked dubiously at me and reluctantly went into the next room. Through the open door I could see the oncologist sitting in front of a computer monitor, laughing and talking with a couple of colleagues. The nurse returned.

'You can go,' was all she said.

Ah, I thought, I have crossed to the other side. I have become just another patient, another old man with prostate cancer, and I knew I had no right to claim that I deserved otherwise.

10

Textbooks tell us that we go through a series of stages as we come to terms with a potentially terminal diagnosis. Stages of disbelief, followed by wild oscillations between terror and denial, clutching at straws, bargaining, anger, despair and perhaps acceptance. This is probably a simplification – everybody is different – but I went through an intense period of blaming myself for having left matters too late. In floods of tears, I cursed myself and apologised over and over again to Kate for my utter stupidity. The one thing I did not do was ask the question 'Why me?' As a doctor, I knew that the answer was very simple: 'Why not?'

I remember lying in bed in the middle of the night, longing to die and wanting to get it all over with, but simultaneously recognising that this was absurd – wanting to die, because I was frightened of dying. And after an hour or two of this I fell fast asleep – I think I had developed compassion fatigue.

And yet this painful process brought some positive revelations as well. I understood that my life was, at the age of seventy, in a sense, complete. I could look back on it and feel that it had been successful. There is nothing more that I need to do. My three children are all approaching middle age and are well and independent.

There are three granddaughters, whom I adore. Much to my regret, however long I live, I will never live long enough to see them reach adulthood. In short, I have fulfilled my biological purpose and evolution now has little, if any, interest in how much longer I live.

I am extraordinarily fortunate in where and when I was born, and to have had the parents and education I had. We are the products of our genes, our culture and our early environment. Success is more about luck than hard work (though hard work is necessary) and is rarely deserved, as having the ability to work hard is itself a question of luck, as is living in a society where hard work of certain kinds is so well rewarded. I have been professionally successful. I have travelled all over the world, have seen mountains, deserts and jungles, many famous cities and have friends in many countries. I have no wish to travel any more. I doubt if future generations – including my granddaughters – will have many of the opportunities that I had. I have made many mistakes and at times clumsily trampled over other people with my enthusiasm and ambition. I wasted too much time and energy in my determination to do everything myself, and too many of the roofs I have built leak when it rains, but I have a loving family and some very good friends. And I understand that the most important reason for living longer is for my wife Kate's sake, and for my family and friends. We are, after all, utterly and completely social creatures. True happiness, I have often thought, is making others happy.

I also understand that future happiness has not yet

happened, and it was pointless to worry that by dying I would miss out on it, or to resent the fact that other people would be enjoying themselves after my death. Besides, I have had my time in the sun – now it is the turn of the next generation, though with global warming the sunlight will no longer be benign.

It is very different if you die young. One of my very first patients when I was a junior doctor was an Irishman in his twenties, Daniel, dying from locally invasive bowel cancer with a 'frozen pelvis' – his bladder and lower intestines infiltrated and incapacitated by disseminated cancer. We were high up on the tenth floor of the Royal Free Hospital, with a fine view looking down on central London from Hampstead Hill.

'All those people down there in the streets,' he said, his voice breaking with anguish and despair, 'why can they go on living and I have to die?'

I had received no teaching or advice as to how to talk to dying patients. I had only been a doctor for a few weeks. I remember very clearly how utterly helpless I felt. I don't know what I said in reply. On the ward rounds the curtains were kept closed around his bed. I would mutter to the professor about the frozen pelvis, and he would nod knowingly, and we would hurry past, although Daniel must surely have heard us through the curtains. We can be so inadvertently cruel. Daniel died a few days later.

I will be missed after my death, but I will miss nothing. When I am lying on my deathbed, I sternly told myself, I do not want to look back on the present time, when I am still relatively well, and feel I wasted it by allowing myself to become miserable at the thought that, sooner or later, I will be lying on my deathbed. I have a duty to my future self to make the most of my life at the moment, both for myself and for others. But I struggled, nevertheless. Wave after wave of despair and anxiety would overwhelm and capsize me, and although I was always able to right myself and struggle to the surface again, there was bound to be another wave on its way. But if I started sinking too deeply into self-pity, I would ask myself what I would think of somebody in my situation, who was thinking the same self-pitying thoughts? And the answer always came back – not much. It was remarkably difficult to pull back in this way and look at myself from the outside. But I usually managed, and it helped.

One of the worst aspects of being a patient is waiting – waiting in drab outpatient waiting areas, waiting for appointments, waiting for the results of tests and scans. When doctors are faced by piles of paperwork and test results (now largely online), it is difficult to keep in mind that each result has an anxious patient attached to it. After my son's successful surgery for a brain tumour when he was only three months old, he had follow-up brain scans for the next ten years. His mother and I learned then about the agony of waiting for results, an agony which few doctors understand until they, or members of their family, fall ill. As for myself, my life will now be punctuated by PSA results every few months. These will tell me if the tumour is starting to grow back – if I have developed 'castrate-resistant prostate cancer', and chemotherapy and the endgame will be started. I hope I am not kept waiting too long for the results.

I also told myself that it would be no bad thing if I died from prostate cancer within the next few years, as it would mean that I would not live long enough to develop the dementia and decrepitude that I fear as much as death. My cancer will be a vaccination against Alzheimer's. And I think of my father who died completely demented, a sad and empty shell. Alas, we remember people as they were at the end of their lives, rather than when they were still truly themselves.

I spent this time mainly in my home in London, either in my workshop making the new dolls' house for Lizzie, or painting picture postcards for Iris and Rosalind. These were illustrations of the fairy stories I was telling them on FaceTime every evening, mainly of dragons and mythical monsters. The illustrations were modelled on medieval illuminated manuscripts, and I even used real shell gold and gold leaf, applied rather badly. As I worked, I listened to classical music on the radio. I derived very real consolation from the thought of how all these composers had died – indeed, there is a countless army of people who have died before me. Death comes to us all, sooner or later, one way or another, I told myself, and is part of life. I will be keeping good company.

But all this was fighting talk, rationalist bravado – I so long not to die. Evolution may have no interest in our living into old age, but it has burdened us with an overwhelming fear of death. It is essential, after all, if our genes are to be successful, that as young parents we fear death and avoid putting our lives at risk, so that our children will thrive. But we carry this fear into old age, when it no longer serves any real purpose, other than to make us miserable when modern medicine can tell us about our death months or years in advance, when we are still quite well.

Yes, death comes to us all, sooner or later, in one way or another, and is part of life, but my wish to go on living is as overwhelming and incontrovertible as love at first sight.

And so I argued and bargained constantly, trying to find reasons why the PSA might be wrong, desperately searching the Internet. This was made all the worse by the fact that it took almost two weeks before I learned the results of the scans that would show whether I had metastatic – secondary – disease or not. The scans – a bone scan, a CT scan and an MRI scan – had been organised, to my great relief, in a matter of days. I was touched by how polite and gentle were the nurses and radiographers. I had enjoyed the scans – lying stretched out in the great machines, a supplicant, naked apart from a white gown and pants, feeling pure and innocent, and hoping that the high technology that was foretelling my death would also save me.

The presence of metastases would probably mean I had only a short time left. But nothing is certain – one of my friends, a professor of neurosurgery in Albania, wrote to me of how President Mitterrand of France lived with advanced prostate cancer for eleven years.

Family and friends assured me that I would be fine, but I thought of all my patients who had been as desperate to live as I was, and whose families had probably said the same, but who had nevertheless died.

At times I was sick with fear, at other times I buoyed myself up by telling myself fairy stories about how the PSA reading was all a mistake, or how my tumour would respond miraculously to treatment. I desperately clung to these fairy stories, finding brief and intense relief in false hope. My mood fluctuated wildly. I suppose this desperate optimism was a form of denial, but denial has a lot to recommend it, as it can bring merciful, albeit ephemeral, relief from living in the shadow of death.

Kate describes my way of dealing with serious problems as 'therapeutic catastrophising'. I conjure up the worst possible scenarios and am filled with terror. I imagined many scenarios of how I might die. I imagined them in clear and horrible detail. I imagined paralysis, I imagined my cold dead body in my bed – I know well enough what corpses look like – and Kate lying next to it, weeping uncontrollably. And I wept bitterly myself at the thought of this scene, and many others like it. Perhaps this was more than just panic and maudlin self-pity, and instead something that helped me accept what might lie ahead, put it to one side, and get on with whatever was left of my life.

The two-week delay before I got to hear as to whether I had metastases or not was extraordinarily unpleasant.

Like most patients I did not dare to contact the hospital to find out. This was partly because I did not want to be seen as a demanding patient, and partly because I was frightened that the scans would show I had metastatic disease. There is always something to be said for living in ignorance. Eventually, in anguish, I asked for my colleague Ken's help, and he contacted the oncologist. I had imagined all sorts of sinister reasons for the silence, but it turned out to be the simple bureaucratic inertia that is so typical of the NHS and is one of the downsides of what Americans call socialised healthcare, where we often have to queue for results and treatment. Although I deplore the application of market economics to healthcare, the profit motive, alas, does seem to make doctors and hospitals respond more quickly– at least, for those who can afford private medical care.

The oncologist rang me two days later.

'I'm very sorry about the delay,' he said. 'My team never told me that you had had the scans. It was a good thing you got in touch to tell me. My team keeps on changing . . .' and he started on a lengthy description of the organisational problems he faced, with which I was all too familiar myself from my days as an NHS consultant. Eventually I had to interrupt him and ask about the scan results.

He told me that they showed no metastatic spread. I felt so relieved by this that I instantly and silently forgave him for not having bothered to chase up the scan results himself.

*

When I went back to the clinic one week later, I saw the oncologist again, after another unsuccessful potty-training session on the urine-flow machine and what I felt were more disapproving looks.

He came in, did not sit down, pushed some papers into my hand, and then stood some distance away from me.

'These are for you,' he said without any explanation. When I looked at them later, I saw that they were the printed reports on my scans.

'Take this to your GP,' he said, handing me a prescription form. 'We need to get your PSA down to under 1 before radiotherapy.'

'What are the chances of that?' I asked.

'90 per cent,' he replied, and my heart leapt. 'But there are risk factors in your case,' he then added.

'My high PSA?'

'Yes, your high PSA'. My heart sank.

He said I needed a biopsy.

'Do I have to have it?' I asked. 'Don't we know the diagnosis already?'

I had often had this argument with my own neuro-oncological colleagues. Biopsies are minor operations to obtain tissue samples for diagnosis, but they are operations nevertheless and are therefore not without risk. That risk has to be justified by possible benefit to the patient. And it is the surgeon doing the biopsy, of course, who must shoulder the blame if the biopsy causes harm, and not the oncologist demanding it. Several of my patients died from biopsies of their brain tumours.

'I'm sorry, but you won't be put into trials of new

cancer drugs if you don't have a biopsy,' he replied. The thought that I might need treatment with experimental drugs was not a happy one, but I bowed to the inevitable.

And that, more or less, was it. I felt as awkward and tongue-tied as on the previous occasion. Did he assume I knew everything about my treatment already, because I was a doctor myself, or that my relative silence meant that I had no questions? Or was any communication to be done by handouts and specialist nurses? Although I was to come to terms fairly quickly with being castrated, it would have been nice to have had some discussion about it. It's not a minor thing to castrate a man, even if he is already seventy years old. I had no clear idea whatsoever as to what lay ahead.

Shortly afterwards 'the team' arrived – a silent radiotherapy nurse who gave me some printed handouts and a friendly 'prostate-specific nurse' who had my urine-flow printout and looked at it disapprovingly.

'Your urine flow is terrible,' she said, in a gently accusatory voice.

'I know,' I replied, feeling anxious and guilty.

'You must drink two and a half litres a day or you will have a terrible time with the radiotherapy. Some of our gentlemen get into a terrible pickle when they have the radiotherapy. We have two months to sort it out.'

Both the nurses looked a little uncomfortable in my presence. Perhaps this was from pity, or perhaps because I was an eminent neurosurgeon – perhaps both. Care, I thought ruefully, seems to be increasingly replaced by printed handouts.

I swore to follow the advice I had been given to the letter. Next morning, I drank a half-litre of water before setting out on my regular run. Completing the two miles became increasingly uncomfortable and I raced back to my home, struggling against overwhelming 'urgency of micturition' as doctors call it. As I turned the key in the lock of the front door, I lost the battle and was hopelessly incontinent. I decided not to drink half a litre of water before my next run. A little research confirmed what I suspected. The need to drink two and a half litres of water every day is nonsense. It is a myth based on a misreading of a recommendation made in a report from the American government in 1946. The report did indeed reckon that on average we need two to two and half litres of fluid intake per day, but added that 30–40 per cent of this water would be in food. In other words, slightly more than a litre of water a day – depending on the outside temperature and your level of activity – is normal and sufficient. I wonder how many other patients in my circumstances torment themselves, obediently following the advice given to them.

When I first met patients with newly diagnosed brain tumours, I would like to think that I spent close to an hour with them, sometimes longer. When I started working in Kathmandu after retiring from full-time work for the NHS, I taught the trainees that an out-patient consultation should always end with the question 'Do you have any questions?' I also told them that this question should not be delivered in a formulaic and hurried manner. It was a point of pride for me to be told that I had already answered all the patient's questions. But it had taken me many years to achieve this, and I am pretty sure that there were many questions the patients thought of afterwards that they had not asked – perhaps like myself too frightened to hear the answer. And when I became a patient myself, I was too shocked and confused to ask much about what lay ahead of me.

11

As I neared seventy years of age, my cancer already present but undiagnosed, it had become increasingly difficult to deny that my body was past its Best Before date. I had started waking in the middle of the night to blunder in the dark to the toilet because of my enlarged prostate. I hated this – especially as sometimes I found it hard to get back to sleep again. I completely blocked out any thought of cancer, although as a doctor I should have known better. I worried instead about the connection between Alzheimer's disease and insomnia, and this would make it even more difficult to get back to sleep. I imagined the amyloid plaques accumulating in my brain and suffocating my nerve cells as I tossed and turned. The arthritis in my knees, which makes driving painful, can also wake me, as do my hands with the dead hand of carpal tunnel syndrome and the painful arthritis in the small joints of the left ring and middle fingers. Some of the joints are becoming a little deformed. I am losing dexterity and am glad that I stopped operating a while ago – that I left too early, and not too late. These nocturnal disturbances would sometimes make me feel, when I finally woke in the morning, that I was being haunted by my own ghost during the night and I wondered about how much longer I might have to live. Perhaps I had some kind of unconscious premonition at this time that I already had cancer.

During the day my neck creaks and crackles and is often painful. It is so stiff that I have to take care not to fall over backwards if I try to look up at the stars at night. My previous retinal haemorrhages mean that I can only see them dimly anyway, and I am amazed at how many stars my granddaughters can see where I see none. I have an artificial lens in my left eye, so that focussing my two eyes is no longer the easy and automatic process that once it was. I have 'paraesthesiae' in my left hand – pins and needles. They come and go, mainly in the little and ring fingers, but sometimes there is a shower of them into all of the hand. I cannot decide whether this is an ulnar nerve syndrome, a C8 radiculopathy, the carpal tunnel problem or even an early myelopathy. In other words, whether I have a trapped nerve in my wrist, my elbow or my neck, or even in the spinal cord itself, the most serious possibility. In cold weather, the tips of my fingers freeze, presumably as a result of deterioration in the micro-circulation of their blood supply. The same deterioration is probably happening in my brain, and perhaps accounts for the ominous changes on my brain scan. My memory is not what it was, and I struggle with simple mental arithmetic. When working out measurements for the elaborate dolls' house I am making in my workshop for my youngest granddaughter, I constantly make mistakes. I often forget familiar names and words, which then suddenly appear a few hours later. It hurts to change the gears on my bicycle as I have arthritis at the base of my thumbs – between the proximal metacarpal bone and the trapezium bone, one of the eight small carpal bones that form the wrist. I have increasing difficulty in swinging my right leg over my bicycle, when mounting and dismounting. It's only a matter of time before I fall over in the attempt. I tire easily and often fall asleep in the afternoon. I find myself sighing, groaning even, with simple physical exertion. In the past, when I was young and ignorant, if I saw this in old people, I thought that it was pathetic and attention-seeking.

Evolutionary anthropologists have convincingly argued that we evolved to dislike exercise. Studies of hunter-gatherers – our forebears – show that they only exerted themselves when they had to, which, of course, they must have done to survive. If we had to walk ten miles to the supermarket every day, we would not feel like going to the gym afterwards. Evolution has shaped us to conserve our energy for the next supermarket visit. I started running regularly in my early fifties. Before then I would only run infrequently, and only short distances, until my daughters pointed out to me that this scarcely counted as exercise at all. So – in order to impress them – I started running more than once round the local park. Each circuit – which I carefully calculated on Google Earth – was two-fifths of a mile, and I gradually increased this to running four to five miles every day. This was after I had met Kate, and I was spending the weekends, when I was not on call, in Oxford, where she lives. The surrounding countryside provided me with wonderful opportunities for running, and over the years I ran longer and longer distances at weekends.

There is a short five-mile run round Christ Church Meadow and then round the University Parks via the path known as Mesopotamia, which runs between the River Cherwell and the millstream for King's Mill. There is a ten-mile run along the Thames, around Port Meadow, and a seventeen-mile run through the villages of Wootton and Cumnor, which involves climbing over a seven-foot deer fence around Wytham Woods (although there is now a new unclimbable deer fence). I would then run through the woods and back to the Thames and the city centre. And then there is a run along the Thames towpath towards Iffley and its perfect Norman church, with pagan centaurs carved in the dog-toothed arch around the South Door. And beyond that towards Sandford and Abingdon – a twenty-mile run there and back which I did once or twice many years ago. In summer the towpath is lined with wildflowers – teasel, chicory, mallow, ragwort, the pretty but invasive alien Indian balsam and many others. Two days after my diagnosis with advanced cancer I went for a long walk in the evening along the towpath, beside the river, sombre in the fading light. I was now unpleasantly aware of the pressure of the tumour in my lower abdomen as I walked, and could no longer dismiss it as something of no importance which I should tolerate, mistaking cowardice for stoicism. I set out in a state of despair and abject misery but returned in the dark, feeling surprisingly calm and resigned to whatever might come.

I am not sure at what point the running, and the press-ups and weightlifting I had taken to doing as well, became painful and an increasing effort, but as I approached seventy, I found that I started to make more and more excuses for exercising less. There were also periods of several months when I had to stop running – a torn hamstring, a proximal hamstring tendonitis, a painful lateral knee ligament. I researched each on the Internet and did not seek the advice of any orthopaedic colleagues. On each occasion I got better, but I no longer enjoyed running. I continued because I briefly felt so well afterwards and also felt reassured that I was keeping old age, cancer, dementia and death at bay. But when I looked at myself in the bathroom mirror, I had to acknowledge that I was looking at a senile caricature of a cover of *Men's Health* magazine, with my sagging, elderly buttocks, crooked neck and wrinkled skin. I resigned myself some time ago when out running to being overtaken by younger people, bounding past me as I stumbled along. There is one particular young woman who often passes me in Christ Church Meadow. She runs effortlessly, springing past me like a gazelle, full of joy – a contrast to my heavy, stumbling steps. As we recognise each other, I smile ruefully, and she laughs kindly in return, and we wish each other a good morning.

And yet I have no wish to be young again. I look back with deep dismay at how driven I was by emotion, and at how little I understood myself. I found it almost impossible to stand back and consider what might be for the best in the long run, both for myself and for others. I was impulsive, tactless and inconsiderate, and on several occasions made a complete fool of myself. There is some neuroscience to this. Self-control, social sensitivity, planning for the future, all come from the frontal lobes of the brain. At least, people with frontal brain damage typically lose these abilities. But the frontal lobes are also the last part of the human brain to become myelinated – for the axons to become insulated. The process is not complete until the late twenties. Teenagers and young adults do not choose to be unreasonable: they cannot help it. There is a particularly sad and unpleasant form of dementia called fronto-temporal dementia where the victims lose all the self-control and consideration for others that comes with the maturation of the frontal lobes. Sometimes, when thinking about climate change and nuclear proliferation, I think the human race has had a collective frontal lobectomy, but this is nonsensical, just as it is nonsensical for me to deride my past self – everything I am now is entirely its creation. Yet looking at global politics and the arms race makes me think of schoolboys in the playground.

There is nothing wrong or unnatural about the various problems I am developing as I grow older. You could almost call them healthy. They signify the wearing out of the body and the winding down of life. The problem, of course, is that they are going to get worse, and there will be some new ones that I do not yet know about. When we are young, we are in ascent, at first in the Garden of Eden, until driven out by the sex hormones of puberty. And then we must find a mate, with all the misery and bliss that that can bring. In middle age we are on a plateau of sorts and kept busy with family and work, but in old age we can only descend, keeping our heads above the metaphysical waters, perhaps, with hobbies and grandchildren, if we are lucky enough to have them. A friend of mine sniffed rather dismissively when I proudly showed him the dolls' house I am building for my granddaughter. He muttered something about 'hobbies' and I felt embarrassed at the way I set so much store by my woodwork. But then I often thought, when still working, that my work as a doctor was only of value because of what my patients did with their lives. If we all became doctors and spent the whole time treating each other, and had no hobbies, the world would be a very boring place.

Growing old calls – at least in principle – for a different mindset from when you are younger. As things can only get worse, you had better make the most of the present, and try to live more in the present, like the hunter-gatherer Pirahã. And indeed, most surveys show people become happier as they enter old age – probably because we need no longer strive and compete, and have come to accept our lot, whatever it might be, until our bodies become too much of a burden and nuisance and start to dominate our lives to the exclusion of everything else. But despite the evidence in front of our eyes – that our bodies are wearing out, or that they are starting to be invaded by cancer – the wish to go on living remains as intense as ever.

*

The initial treatment of locally invasive prostate cancer is grandly called 'androgen deprivation therapy' (ADT) in the literature – a euphemism for chemical castration. Prostate cancer thrives on the male hormone testosterone, and supressing testosterone usually, but not always, causes the tumour to shrink. After my diagnosis I inevitably turned to the Internet to learn about chemical castration. When I was a medical student, I remember a retired GP with prostate cancer coming to talk to us about the awful effects of the treatment. The drug used then was stilboestrol (pronounced stil*beast*rol), the same beastly drug that Alan Turing was given in lieu of being sent to prison for homosexuality. It may well have contributed to his suicide. There have been recent reports of the drug being used surreptitiously by wives in China to bring their erring husbands to heel – apparently you can buy it from vets.

So by blocking testosterone production the tumour will regress for a while, although always recurs sooner or later, without further treatment. This effect was discovered by the American researcher Charles Huggins in the 1940s, who won the Nobel Prize for finding it The connection between the prostate and the testes had been noted as early as the late eighteenth century by the anatomist John Hunter, who observed that castrating dogs resulted in their prostate glands shrinking. At first, irreversible surgical castration was used for treating prostate cancer. I remember many years ago, when working in general surgery, seeing ping-pong-ball-like prostheses inserted into men who disliked having an empty scrotum.

Castration is now achieved chemically, using a variety of drugs that act on the brain's regulation of testosterone production, or on the body's receptors for testosterone, with effects that are in principle reversible.

As with all hormone production in the body, there is a complex system of checks and balances, orchestrated by the brain, involved in the regulation of testosterone. Testosterone from the testes (and to a much lesser extent from the adrenal glands) circulates in the blood, crosses the blood brain barrier and binds to receptors in part of the brain called the hypothalamus. In response the hypothalamus produces a hormone – luteinising hormone-release hormone (LHRH) – that in turn acts on the pituitary gland, which responds by producing luteinising hormone, which causes the testes to produce testosterone. The mainstays of modern treatment of advanced prostate cancer are LHRH 'agonists' that mimic the action of LHRH, resulting in the 'downregulation' of the LHRH receptors in the hypothalamus. The brain, so to speak, thinks that it is awash in LHRH and so stops producing it, when it is in fact awash in a man-made chemical, that does not stimulate the pituitary gland to produce luteinising hormone. As a result, the testes stop producing testosterone.

Chemical castration causes prostate cancers to become smaller. The tumour is starved of the testosterone on which it needs to grow. After castration the cytoplasm of the tumour cells – the fluid that fills the cells – is described by pathologists, looking down their microscopes, as becoming 'foamy' and the cells' nuclei, that contain the DNA, as shrunken or 'pyknotic'. But the cancer does not die, and it is only a matter of time before it starts growing again. The cancer is a living thing, but with its constituent cells each subtly different and in competition with each other. By a process of Darwinian evolution, the cells that do not need testosterone flourish, and will start to predominate, while those that do need testosterone will wither. When this happens, the patient is said to have developed 'castrate-resistant prostate cancer', which is revealed by the PSA starting to rise again. Further treatment with chemotherapy can then be started, which poisons the cancer cells as opposed to starving them, but this will only temporarily slow the tumour down as cells that are resistant to the chemotherapy will start to predominate. It is now the endgame – sooner or later, the cancer will spread to the rest of the body and kill the patient. And it may cause mayhem locally as well. How long this takes will also depend on the complex biochemical environment in which the tumour lives – some parts of the body will support its growth, others will not. This explains why some cancers – such as breast and lung cancers – have a predilection to spread to the brain and liver while other cancers, such as mine, flourish in bone.

Of course, the patient may die from something else first, given that prostate cancer is mainly a disease of old men. This makes it very difficult to quote mortality rates for prostate cancer. Did the patient die with or from the disease? It took me quite a while to work out from reading the professional literature on the Internet – given how little I had been told at the cancer hospital – that the risk of biochemical recurrence in my case within five years, given my high PSA and tumour grade, is 75 per cent. What this means in terms of life expectancy is impossible to know – but as I am otherwise fit and healthy and unlikely to die from another disease within the next few years, it seems probable that prostate cancer will be my final disease. Sometimes I think I am lucky to have this insight into the future, and the concentration of the mind that it brings. But not often.

After a year of chemical castration for my prostate cancer, I found exercise more and more of an effort. I would dread it and find constant excuses for postponing it. I don't know whether this was because the lack of testosterone in my body weakened my willpower or my muscles, or perhaps both. I enjoyed it less and less, but nevertheless the feeling of profound relaxation and mental clarity that came afterwards continued, and it was this, as well as fear and loathing of old age and the effects of hormone therapy, that kept me at it. I am not fighting cancer, but only myself and the side effects of the treatment. This feeling of well-being after exercise is now attributed to the endocannabinoid system in the brain, a system of neurotransmitters and enzymes that are mimicked by marijuana. Like so much about the brain, it is not fully understood. Research has shown, apparently, that this feeling only comes to people who exercise regularly and for at least twenty minutes every day. Serious runners talk of a 'runner's high' which I have rarely, if ever, experienced while running. It is said you only experience this if you exercise continuously for more than three to four hours and is related to endorphin release in the brain, chemicals which are mimicked by opiate drugs. But the neuroscience of this again remains obscure – perhaps the endorphins are released because running for such a long time becomes painful. What is clear is that we find exercise difficult, but the after-effects, with practice, very enjoyable. I find it strange how the present weighs so heavily on us, and that it is so difficult to sacrifice present comfort for a much greater benefit in the future, just as I find it increasingly difficult to get out of bed in the mornings.

12

I do not want to die – but then, who does? But nor, to state the obvious, do I want to become old and decrepit. It used to be thought that human life came with a fairly specific time limit and that there was something 'natural' about dying in your seventies, but trying to arrest or even reverse ageing is now serious science, and no longer the exclusive preserve of cranks or quacks. We are part of Nature, and everything we do – including all the technology that is so much a part of human life – is as natural as sex and trees. It is no good saying that trying to extend human life is against Nature, though whether it is sensible or not is a different matter. I find the idea of extending human lifespan appalling but can't deny it is a prejudice on my part, which I must overcome, if only to understand the science behind it.

There are enormous differences in the lifespans of different creatures – from insects that might live a matter of hours to Greenland sharks and bowhead whales that can live for hundreds of years. Some – like the Galapagos tortoise – show few signs of ageing even in old age. There are various theories as to why we age, but they all must be accounted for in terms of evolution and natural selection, in accordance with Theodosius Dobzhansky's famous dictum in 1973 that 'nothing in biology makes sense except in the light of evolution'.

One of the main theories of ageing is 'antagonistic pleiotropy' – in simple terms, of evolutionary neglect. It is now understood that the same gene can have different effects – pleiotropy – in different circumstances. A gene which increases reproductive success in youth may have a damaging effect in later life. But this gene will be selected for and spread, despite resulting in cellular decay later in the organism's life. Evolution by natural selection is simply a mechanism, but I am unable to avoid the pathetic fallacy of attributing purpose and intent to it. Natural selection has no interest in the sufferings of old age. It has abandoned me.

There are exceptions to this – there always are: fish, for instance, and, to some extent, humans, as we live much longer than other primates. Women live well beyond their reproductive age. The most plausible explanation for this phenomenon in humans is the so-called grandmother hypothesis. It seems that humans are unique (although there is evidence that whales might be an exception) in the way that grandmothers are involved in the rearing of the next generation, and it is suggested that this explains our long lifespan compared to other primates. Grandmothers enable their daughters to have more babies and so genes that prolong a grandmother's life will be relatively successful compared to those that don't.

Reproduction is not just about having babies, but equally about childcare, and doing our best to see that our children and grandchildren reach reproductive life themselves. If we do not take care to avoid danger and possible death, our children and our genes will not survive. But now that I am entering old age, I am lumbered with the same fear of death that helped my genes succeed when I was young. But my fear of death is now pointless. I do not believe in an afterlife, and it makes no sense to be afraid of nothing. And although my fear of death now takes the form of fear of *dying*, I suspect that beneath this there is still a deep, irrational fear of death itself, of nothingness, of having no future. So much of present time is spent thinking about future time – either with anxiety or happy anticipation – that the thought of having only the present left, with no future to come, is frightening.

Driven by this fear of having no future, enthusiasts for life extension, who call themselves transhumanists, see the genetic basis of lifespan as grounds for great optimism that death will be overcome. Death, they argue, is no longer inevitable – just as a salamander can grow a new limb after amputation, or a lizard a new tail, it should be possible, with suitable technological advances, for us to regenerate ourselves endlessly. They say that the genes that determine how long we live will be hacked, and the cellular processes they control altered to keep death at bay. Elderly, anxious billionaires are funding research into ageing and death, in the hope that they can avoid it, inspired by their pathetic belief that living longer is what matters most in life.

I find it very difficult to comprehend that I am

trillions of cells (and even more bacteria, sheltering in my gut, upon whom I depend). These cells are all the descendants of the original first cell, with each generation becoming more and more specialised – into at least 200 different types of cells – liver cells, skin cells, brain cells and so on. Although each cell contains the DNA for forming my entire body, early in development each cell is programmed to express ony a certain set of genes, appropriate to its specialised role. It is not just our bodies that are sculpted by natural selection, but all the trillions of individual cells. Each cell is driven to live and to reproduce itself, but also to co-operate and not compete with all the other cells with which it lives. It has to be held in check by many complex mechanisms to ensure that it remains only part of a whole greater than itself. Failure, of course, is cancer, which – in a perverse sort of way – is a fatal expression of the force to live.

The Nobel Prize-winning stem-cell scientist Shinya Yamanaka showed that almost all specialised cells can be reprogrammed backwards to become what are called induced pluripotent cells. These are cells very similar to the original precursor stem cells that produced the specialised cells that Yamanaka started with. Enthusiasts for life extension see Yamanaka's technique as the route to eternal life. Research to date in mice, using this technology, tends to produce tumours called teratomas instead of long-lived mice, but we are only at the beginning of this kind of research. There are recent reports which suggest this problem can be overcome. Teratomas can occur spontaneously in humans, particularly in children, and sometimes in the brain. I remember once finding parts of a minute ribcage in a teratoma I removed from a child's brain. There is a report of a miniature brain and skull found in a teenage girl's ovaries. Nature does sometimes get it terribly wrong. Yet it is miraculous how most of the time stable, reproducible life forms arise from so much complexity.

Cancer can develop in almost all cells, and almost all complex creatures. It is primarily a disease of old age and caused by the accumulation of random mutations in DNA which are not repaired. Yet there is a fascinating fact, first noted by the epidemiologist Richard Peto – and hence called Peto's Paradox – that large, long-lived animals such as whales do not have higher rates of cancer than smaller animals with shorter lives. The cells of a whale are no larger than the cells of a mouse, so whales have many millions more cells than mice, and yet whales do not suffer more cancer than smaller animals. Indeed, they suffer less cancer. There are many theories that attempt to explain this paradox, and the enthusiasts for life extension see it as evidence that there is nothing fixed about lifespan, at least as regards cancer. The problem, however, is that the mechanisms for suppressing cancer seem, in some way, to be closely related to ageing – that ageing is the price we pay for avoiding cancer when we are young and busy reproducing ourselves.

Is a long life a better life? Transhumanists say that at a conservative estimate we soon will all live to 150 (the really enthusiastic say to 1,000). I have little sympathy with this wish to prolong life, just as I have little sympathy with plans to colonise Mars, when we are busy wrecking our own planet. The vast sums of money involved could be better spent elsewhere. The universe was perfectly happy before bipedal linguistic apes with opposable thumbs evolved on Planet Earth. It is now understood that there are many billions of stars in our own galaxy. It is also understood that there are many billions of galaxies beyond the Milky Way. Each in turn with billions of stars, and most, in all probability, with planets. So it must be unlikely that intelligent life is confined only to our planet. The history of science is largely a history of the refutation of human exceptionalism – the earth is not the centre of the universe; human beings are animals. As the great zoologist J. Z. Young observed, we are risen apes, not fallen angels. Besides, what is so good about intelligence? The wish to colonise space and take our intelligence beyond the earth's boundaries is just the deep human desire to spread and propagate, shared by all living things, and at which we have been especially successful, to the cost of so many other life forms. But this success has made many of us see ourselves as more important than other life forms. If there is anything good about our intelligence, it is that it gives us the ability to love and respect all forms of life, and not just our own. I am not greatly troubled by the idea of the human race coming to an end – in the very long term this is inevitable, after all. As the philosopher David Hume observed on his deathbed, the thought of his no longer existing once he had died, bothered him no more than the thought that he had not existed before he was born. But I am appalled by the suffering that the decline and end of the human race will probably involve, and I think about my granddaughters and their possible descendants, and climate change, and all that it will bring in its wake.

Although most cosmologists reckon that the universe will eventually come to an end, it is an incomprehensibly long time away. Some of them have come up with the disconcerting idea of Boltzmann brains. The Second Law of Thermodynamics (developed by Ludwig Boltzmann in the latter part of the nineteenth century) explains thermodynamics – the study of matter, energy and heat – in terms of statistical probability. There is no law which says that the trillions of gas atoms flying about randomly in the room I am sitting in cannot suddenly rush out of the window and leave me gasping for air. It is just highly unlikely. If the universe came into existence as a highly improbable random fluctuation of some kind (don't ask me what kind!), it must be less improbable for simpler structures such as the Milky Way to come into existence. And even less improbable for the solar system and, so the argument goes, even less improbable that billions of particles might suddenly assemble themselves into a brain. So perhaps I am a Boltzmann brain, and what I think is the real world is just a pattern of electrochemical impulses in my brain, formed by randomly self-assembled particles of matter. Which, in a sense, it is, whether you believe in the possibility of Boltzmann brains or not.

There is, however, a problem with life extension for

sceptics like me. There is good evidence that increasing lifespan – at least in mice and worms – can prolong not just life but healthy life, healthspan as well as lifespan. Ageing is a disease, with specific mechanisms, and there is nothing divinely ordained about the deterioration of our bodies. How could I reject technology that might lessen all the sorrows of ageing – arthritis, cataracts and macular degeneration, cancer, osteoporosis and the withering of my brain seen on my brain scan? And yet I find the thought of a world populated by more and more old people, however fit and healthy they might be, horrible. This is, of course, happening already, with the great demographic transition of declining birth rates, brought about by the emancipation of women, birth control, sanitation, vaccination and declining infant mortality. But even if human beings – though perhaps only very wealthy ones – are soon all living to 150, will their suffering at the end be any less, just because they lived so many extra years? Will our lives be any more meaningful, just because we have managed to postpone death? And will the future belong to the old, when it really should belong to the young?

Just as pleiotropic genes do different things in different circumstances, there are few, if any, characteristics in living creatures that are determined by single genes. After the rather shocking discovery that there are only 20,000 human genes, it came to be understood that genes work together in very complicated ways – that the colour of our eyes, our height, our intelligence, are polygenic and determined by multiple genes. Hundreds of genes, for instance, have been implicated in schizophrenia. Altering one gene with one of the new gene-editing technologies such as CRISPR might have all manner of unpredictable, undesirable effects downstream. Genetic modification in food crops is one thing – and a vital tool for combating climate change and environmental degradation. Altering a human genome for certain rare, single-gene diseases is a different matter from editing normal genes in the hope of improving the body they produce. It is unlikely, therefore, that there will be any simple genetic hacks to stop us growing old. And even if there are, it is quite likely that there will be all manner of unwanted side effects, of which cancer is only one, just as my hormone therapy has so many side effects. Or perhaps we will live for hundreds of years, but at the speed of a tortoise.

Dietary restriction has been shown to increase significantly the lifespan and healthspan in mice, other small rodents and non-mammalian creatures. But apart from the fact that a life of severe caloric restriction might rather reduce the pleasure of being alive, the problem remains that what works in mice often fails to work in people. We are not, it has been observed, large mice. And scaling up such experiments into people will pose all sorts of ethical problems, which may well be insurmountable.

But the research will continue, funded by anxious, ageing billionaires, who are driven by the selfish, hedonistic greed that is destroying so much of life on our planet. You can never predict the future, but I hope that the genetics of ageing prove to be impossibly complicated and life extension is merely a venture-capitalist's dream, or at least that it is a long way away. I certainly will not live to see it. You can never know what the researchers will find, but if they are eventually successful I doubt if living longer will make the billionaires' lives any more meaningful, or their dying any better.

13

My biopsy was carried out under general anaesthetic as a day case. The staff could not have been nicer. I was welcomed by name at the entrance to the Day Case Surgery Unit by a friendly staff nurse, who did not know that I was a surgeon. The whole procedure was choreographed perfectly – from my being checked in, to undressing and being clad in gown and skimpy throw-away knickers, to meeting the anaesthetist and surgeon, to being finally walked down to the operating theatre.

I enjoy general anaesthetics – they hold no fears for me and are miraculous. You go out like a light and come round later in a pleasant haze of painkillers.

'Are you in pain?' I remember the recovery nurse asking me, as I floated upwards into consciousness. 'What is your pain on a score of one to ten?'

'Nine,' I replied, thinking it best to exaggerate – something we all should do in these circumstances. So I was given fentanyl, and experienced only mild discomfort as she pulled out my urinary catheter. I have forgotten the nurse's name, but she told me she came from Nigeria and that she was also studying business management at the University of Westminster.

Based on work using tadpoles and olive oil, early researchers into general anaesthetics, most of which are fat-soluble, suggested general anaesthetics worked by interfering with the membranes of all the nerve cells in the brain. This is no longer believed – more recent work using genetically modified mice has shown that general anaesthetics depress activity at synapses that have an excitatory role and enhance activity at inhibitory synapses. They depress the activity of nerve cells by acting on the 'ion gates' in cell membranes, that are controlled by neurotransmitters. These ion gates control the electrical state of nerve cells by pumping ions – particles with an electrical charge – in or out of the nerve cell. The molecular mechanisms involved are increasingly well understood. And yet, as with so much neuroscientific research, how it all fits together is not well understood. But it does seem that general anaesthetics target specific areas of the central nervous system, and not all nerve cells indiscriminately.

General anaesthesia involves much more than just loss of consciousness. Immobility, loss of memory, excitation and agitation, muscle relaxation and paralysis of breathing can all occur, depending on the dose of the anaesthetic. When I was a medical student, we were shown an experiment on an unfortunate cat, in a glass box. It was anaesthetised (with nitrous oxide, I think). It passed through the four textbook stages of anaesthesia – analgesia, delirium, surgical anaesthesia and finally, respiratory arrest. And so the cat was dead at the end of the experiment, as it was not put on a ventilator, which would happen with a human patient. I watched this with a mixture of fascination and dismay. As a way of teaching the principles of general anaesthesia, I was not sure if this was any more effective than a lecture or reading a book, and I felt sorry for the cat.

It is now thought that each of these effects results from the action of general anaesthesia on different parts of the central nervous system – there is no single, central switch being turned off.

It is easy enough to conceptualise what you might call the physical effects of anaesthesia – the paralysis, the loss of memory and agitation – but the central mystery of consciousness remains. How does consciousness arise, and from where? Despite all the research and writing, we really have very little idea. And this is rather extraordinary when you stop to think about it – that we understand so little about the most important aspect of being alive.

You learn little, if anything, about consciousness from introspection. To examine your own consciousness is like a snake swallowing itself, or, as the 'father' of psychology William James wrote, like turning up the light to examine the dark. But if you accept that consciousness is generated by the activity of nerve cells, you will immediately realise that something remarkable is going on. The electrical impulses with which nerve cells influence each other take time to travel across the brain from cell to cell, and yet we experience an intense feeling of unity and 'nowness' when we are awake and conscious. William James talked of the 'stream of consciousness'. It is easy to think of our vision as a series of still photographs flashed quickly past, as with cine film, creating the illusion of movement, but this metaphor breaks down when you think about sound. How can there be an auditory equivalent of a still photograph? Our conscious experiences are *constructed* from the electro-chemical dance of our nerve cells, and this takes time and space. Admittedly, a few milliseconds and millimetres – but our feeling of an immediate, singular present is clearly an illusion.

Given that nerve cells act over space and time, perhaps we should not be surprised that scientific experiments show some strange effects. Some of the earliest, and most famous, were Benjamin Libet's experiments in America in the 1980s. These showed (and have been confirmed subsequently many times) that the conscious decision to move your hand comes a few milliseconds *after* electrical activation of the brain's hand area, as revealed by electrodes placed over the scalp, which record the electric currents on the surface of the brain beneath. Some critics argue that moving one's hand is a small and unimportant part of cerebral activity but, if you think about it, any 'conscious' decision must come from somewhere – it cannot just arise from nothing. That is, not unless you are a 'dualist' and like Descartes believe in an immaterial human soul or self. Libet's experiments have inevitably added to the already excessive literature about 'free will'. Free will was a concept introduced by Catholic theologians to explain how evil arises in the world, despite a benign God. Although God is omnipotent, all bad things are our own fault.

Studies have shown that a remarkably high proportion

of violent offenders in prison have a history of previous head injury, or an abusive childhood, or both. In my work I often saw patients with physical brain damage, typically to the frontal lobes from accidental trauma. This often causes personality change, almost invariably for the worse. They were almost always unaware of this themselves, but it could be quite terrible for their families. Such experiences make it very hard indeed to believe that thought and feeling, and consciousness itself, are not generated by physical processes in our brains. Free will might be a legal necessity for an ordered society, but it is an illusion. Our decisions are determined by our past. It is an illusion, just like the illusion that the pain in my little finger – which I cut badly a few days ago on a mandolin food slicer – is in my finger, when in reality it is a complex pattern of nerve impulses in my brain. But this doesn't mean that pain isn't painful, or that difficult decisions are not difficult. It just means we don't understand how our brains work.

But there is some scientific understanding now – as opposed to mere philosophical or psychoanalytic speculation – of the relationship of the unconscious to the conscious, but as with so much science, answering one question just opens the door to a room with yet more doors.

The history of scientific progress is a complicated dance between new technology, new ideas, and resistance to them. It has not always been a history of linear progress.

The so-called scientific revolution of the seventeenth century was inextricably tied up with the invention of telescopes and microscopes in Holland in the early seventeenth century. For the first time it could be seen that all sorts of things existed beyond the scale of human vision – that our everyday, commonsense view of the world was incomplete. Yet magnifying lenses were known to the ancient Romans, and Antonie van Leeuwenhoek's discovery of bacteria in the seventeenth century had to be rediscovered 200 years later. Equally extraordinary is that Paracelsus was using ether to anaesthetise chickens in the sixteenth century, but it took 300 years before it was used on humans, and medicine was transformed.

Experimental psychology in the past had little choice other than to regard consciousness and the inner workings of the brain as a 'black box'. Only inputs and outputs – stimulus and behavioural response – could be fit subjects for objective scientific study. Technology has changed all this, and we can indeed now peer into the black box and gain some understanding of the inner workings of our brains. But it is a strictly limited view – EEG recordings summarise the activity of billions of nerve cells, and only on the brain's surface. The temporal and spatial resolution of MRI and functional MRI (fMRI) are also limited. One cubic millimetre of cerebral cortex can contain up to 100,000 nerve cells and a billion connections, which operate in milliseconds, whereas fMRI has a temporal resolution of three to four seconds. The most powerful non-functional MRI scanners have a spatial resolution of slightly under a millimetre and functional MRI is several times less accurate.

Electrodes can be put inside the brain and even into individual brain cells. This is infinitely more accurate than EEG, but for obvious ethical reasons, such research is very limited. And bear in mind that there are some 86 billion nerve cells in our brains. All this amazing technology is still a little like looking at the starry sky at night through a pair of cheap, low-powered binoculars. So there are very real limits to what present technology can tell us. It's anybody's guess whether future technologies will change this.

There is a disconcerting experimental technique which cognitive scientists call 'masking'. It has proved a vital tool in investigating consciousness. If an image is flashed up on a computer screen for a few milliseconds, we will easily see it. If, however, a second unrelated image – the 'mask' – is flashed up almost immediately after the original image, we will not *consciously* see the first image although it must still be present somewhere in our brain.

At first, this seems extraordinary – how can something that occurs *after* an event obliterate that event from our conscious perception? But remember that once we accept that everything we think and feel is produced by the activity of nerve cells, and that it takes time for nerve cells to communicate with each other, we should not be especially surprised by this apparent paradox. Indeed, research has shown our conscious perceptions lag about one-third of a second behind any initial stimulus. It is as though our conscious self acts as a reporter, providing some kind of executive summary of what is going on in the unconscious. The 'mask' can disrupt the reporting, so that the original image remains in the unconscious.

Using masking methods and technology, images can be shown to move in and out of conscious perception, and yet be present in the brain at an unconscious level. And it is not just visual images – it can be shown that emotional reactions (in the amygdala) can be present without conscious awareness of them. Using EEG, an unconscious perception or emotion can be seen to fizzle out quickly, but if the perception becomes conscious, it spreads rapidly throughout much of the brain. A visual image is first registered in the brain at an unconscious level in the visual cortex. If it becomes conscious, the initial unconscious electrical activity spreads rapidly and widely over the brain. It spreads in both time – for more than 300 milliseconds – and space, especially to the frontal and parietal lobes beyond the visual cortex. Consciousness is therefore to be seen as the electrical activation of broad areas of the cerebral cortex by an initially unconscious stimulus. Researchers use a variety of metaphors to describe what occurs when a perception becomes conscious. There is an avalanche, a phase transition, a tsunami, a process analogous to the collapse of the wave function in quantum mechanics, a rapid ramping up. According to Stanislas Dehaene, one of the leading researchers in this field, the conscious self is like the chief executive of a large corporation with thousands of employees.

This multiplicity of metaphors once again shows us just how problematic is the nature of the conscious and unconscious. They are not separate entities but parts of a whole. It must be very uncertain as to whether we will ever be able to conduct experiments on brains equivalent to the physicists' experiments on matter, which can then be modelled and explained in mathematical terms. We cannot take our brain apart, test out different bits and then put it back together again.

There are two ways of looking at the conundrum of how conscious experience arises from physical matter. One is the view taken by many, but not all, neuroscientists. Consciousness is simply a property of nerve cells that emerges when they are connected in certain configurations. To search for some underlying essence is like the futile search in the past for some essence of life, or for phlogiston, an essence of heat. We now know that heat is simply the motion of atoms and life is simply the self-replication of certain molecular configurations. And yet this analogy is misleading. Heat is the word we use to measure the movement – the kinetic energy – of atoms. The word also describes the perception in our brains of this energy, which itself is a pattern of physical activity in our nerve cells. So it is not really the case that there is no such thing as heat. As for life, this is simply a question of definition. It is hard to know whether viruses should be called alive or not, as they can only reproduce themselves by hijacking cells that have the machinery for transcribing DNA or RNA into proteins, which the viruses themselves lack.

If you take this a step further and believe that consciousness is an emergent property of information processing per se, it follows that consciousness could arise in computers, a difficult idea to swallow, though impossible to disprove. Brains are physical systems and computers are physical systems, and they process information, but the similarity ends there. There are profound differences in their structures and components, and how they work. As it is, we understand remarkably little about how our brains work. It is not even certain that brains compute, using algorithms and a code in the way that computers do. The word intelligence is hard to define, and we have no idea as to its neurological basis. The very phrase 'artificial intelligence' is misleading – the 'intelligence' of an AI is utterly different from the intelligence of, say, a child. A child only needs to see a cat once before she can identify all cats in future, whereas AIs need to 'see' millions of pictures of cats before they can identify them. Nor do AIs 'see' in the way that we do. So-called 'adversarial attacks', where only a few pixels in an image are altered, can cause an AI to be hopelessly wrong, even though to a human eye the image is entirely unchanged.

I am not at all worried about super-intelligent AIs in some way replacing us in the foreseeable future, although it may well render large numbers of people unemployed. (Most economists argue that this does not usually happen with the introduction of new labour-saving technology, and instead news jobs are created.) But I am deeply worried about the way AI will be used in a malign way by my fellow human beings, and especially by despotic governments. The loss of anonymity, for instance, that comes with omnipresent CCTV and facial-recognition software puts enormous power in the hands of a few people.

The idea that brains and computers are essentially similar leads to all sorts of delightful cyberpunk fantasies about uploading our brains onto computers. And this leads to the further problem of how we could tell whether a computer was conscious or not. This has been a field day for the philosophers, who write at great length about these problems, but I struggle to understand them. I cannot escape the feeling that much of this is just playing at word games. But nor can I quite escape the fear that I am simply not clever enough to understand philosophy. I did, after all, abandon it as a student and now, in old age, I wonder whether the problem is made even worse by the white-matter hyperintensities on my brain scan.

The other way of looking at consciousness – taken, for instance, by the mathematical physicist Sir Roger Penrose – is that our understanding of the physical world is incomplete. If thought and feeling are created by our brains, they are part of the physical world, and our brains must be subject to the laws of physics. But physics has nothing to say about consciousness and is therefore incomplete. There is a similar incompleteness, he argues, with theoretical physics which has failed, as yet, to combine quantum mechanics and general relativity, and cosmologists struggle with dark matter and dark energy – dark because we do not understand them.

There could be, in other words, a deeper level of explanation of physical phenomena which we currently do not understand, and perhaps never will. As the evolutionary biologist J. B. S. Haldane observed, it's not just that the universe might be stranger than we think, but that it might be stranger than we can think. This view also can lead to other sorts of fantasies, about 'higher levels of quantum consciousness' – as opposed to the ones about emulating our brains on computers. What both kinds of fantasies have in common is our underlying fear of death and our longing for eternal life.

But both views of consciousness, and their accompanying theories, come to grief on the simple fact that consciousness is subjective. It's not an easy subject for scientific study beyond simple description – perhaps even impossible. As a thought experiment, I sometimes try to design a machine which would enable me to study my own consciousness – I have made no progress to date.

I had been reluctant to undergo the biopsy as it can cause post-operative acute urinary retention. This is an emergency, although easily dealt with by inserting a catheter. I dreaded this, so when I found I could pass urine in the toilet when I was back in the Day Case Unit, I felt very relieved.

'Did you go OK?' the friendly staff nurse, who was sitting at the nurse's station, called out as I emerged from the toilet.

'Yes,' I replied happily. 'It came out at an odd angle, but I have mopped up the floor.'

Once back home, I wrote to the hospital matron, congratulating her on how well I had been treated.

I had one further radioactive scan later that day – a PET CT scan which would show if the cancer had already spread to my lymph nodes – and then was collected by Kate and my neighbour Selwyn and taken home. While waiting for them to arrive, I walked along the Fulham Road. Although the radiation dose from the scan was minute, there was a bio-hazard sticker round my wrist, which I proudly wore like a soldier's wound stripe. Although the second lockdown had started, I found an art shop that was open – what was so essential about what it sold, I do not know – but I bought some paintbrushes for working on Iris's and Rosalind's picture postcards.

And that evening I took the first of the hormone therapy pills.

I inevitably started googling the side effects of chemical castration and survival rates for advanced prostate cancer. I studied the graphs and tables in the professional literature, showing mortality from prostate cancer, trying to read my future. But it was little better than trying to read my future in the stars – it did not show what would happen to *me*, it was only statistics, only probability. I became quite irrational – some things I read filled me with dread, and I was convinced that soon I would be dead, others made me briefly and wildly hopeful.

By all means, I would tell my patients, google your

illness, but take care. It can be frightening, it's not all accurate and it's only probabilities, never certainty. I told them I would try to tell them everything that I thought they needed to know. But now that I had been diagnosed with cancer myself, I understood that my knowledge of what they needed to know had been rather limited.

All drugs have side effects. The list of side effects of castration is very long. You realise quickly that these lists are not really there to help patients but to protect doctors and pharmaceutical companies from complaints and litigation. At least some of the websites distinguished common from uncommon side effects, and reassuringly added that nobody suffered from *all* of the side effects. Some of the side effects were very unappealing, such as gynaecomastia – breast development (which I started to develop very slightly after a year of treatment). Weight gain around the waist and loss of body hair are probably the commonest side effects, as well as impotence and loss of interest in sex, muscle loss, osteoporosis and bone fractures. The list does not end there. After one year of hormone therapy I have come to dislike seeing myself in the bathroom mirror – I have acquired the plump and hairless body of a eunuch, and look rather like an outsize, geriatric baby. Vanity, of course.

Many of the listed side effects – such as headaches, dizziness and chronic fatigue, constipation, diarrhoea – are very non-specific. The problem with these non-specific symptoms is that we are all deeply suggestible. It is called the nocebo effect. This is the opposite of the benign placebo effect, where people will feel better because they have been told to expect to, and not because of a direct effect of the particular treatment they have been given. With the nocebo effect you feel worse, because you expect to. Looking at my brain scan had had the same effect – it took a while before I could overcome the feeling that I was already dementing. And depression was also listed – but what man will not be depressed by castration and the possibility of imminent death? Is that a side effect?

Testosterone is popularly thought to be responsible for aggression in males, but research does not bear this out. It seems that if a man is in a social situation where co-operation is valued, testosterone will make him more co-operative; if in an aggressive situation, he will be more aggressive, especially if he was already prone to aggression. Most experiments along these lines have been on WEIRD psychology students – WEIRD being western, educated, industrialised, rich and democratic – so it is hard to know just how much they mean. The literature on the psychological and cognitive effects of hormone therapy is inconclusive. Perhaps I became less competitive with chemical castration, but after retirement I have only had my past self with which to compete.

There are various treatments for some of the side effects of chemical castration, usually with impressively long names. Megestrol acetate and medroxyprogesterone acetate for hot flushes and phosphodiesterase inhibitors and vacuum devices for impotence. However, as one article I read observed, 'lack of libido often limits patients' enthusiasm for pursuing treatment to restore erections'.

As for myself, I do not miss my libido or erections. Indeed, in many ways I am glad to be free from them, especially when I think of all the misery they had caused me in adolescence and middle age, and the madness – which feels divine but is often absurd – that comes with falling in love. But I have to admit they had also been responsible for the greatest joy in my life – of having a family. When I was seen in the radiotherapy clinic by a kind and sympathetic nurse, halfway through my month of radiotherapy treatment, who took me through a long checklist of the symptoms from which I might be suffering, I joked that the world would be a better place if all middle-aged men were on ADT. I thought of the Chinese wives who have been slipping stilboestrol to their unfaithful husbands. But if I had been on ADT, I reflected afterwards, I would not have been unfaithful, my first marriage would probably not have ended and I would not have subsequently met and married Kate, who transformed my life, and made me a better person. Besides, my first wife is much happier as well, and we are good friends once again. So for all this I must thank testosterone.

You can make yourself very unhappy by reading about your medical condition on the Internet. Although it is often very useful, it is a poor substitute for a sympathetic doctor, who can guide you through the thicket of statistics and give you some hope despite them. It reminded me of my early days as a medical student, when I started to learn about many horrible diseases that all started with quite modest symptoms. Like most medical students, I went through a brief period where I was frightened that I had all sorts of deadly diseases, but I soon learned that illness only happened to patients and not to doctors.

Hope from a kind and reassuring doctor feels quite different to hope from a web page or printout. It is not necessarily that, childlike, you expect the doctor to cure you. Instead, it is the reassurance that the doctor cares for you, and will try his or her best, even though ultimately in a case like mine they will probably fail. The medical author Gavin Francis, whom I greatly admire, describes his role as a GP as being 'a guide through the landscape of illness'. Much of this support in the NHS is now provided by specialist nurses, printouts and questionnaires, and not by doctors. I was asked to fill in three 'holistic healthcare questionnaires'. The nurses I met were all excellent – kind and sympathetic; but they are not responsible for the decisions that have been made about your treatment. You are unable to discuss the decision-making in your case, only the details of its application and side effects. This has a very disempowering effect. But at least this disengagement from patients must make the oncologist's life considerably less stressful.

It is very difficult to derive hope from a web page – you are easily overwhelmed by panic and despair. After a while I stopped reading obsessively about prostate cancer and decided that I should just get on with living whatever life I had left. The only certainties in life, as is often pointed out, are death and taxation. I became more adept at managing my thoughts. If I started feeling a wave of anxiety approaching me, I told myself that I would receive appropriate treatment, that we all have to die sooner or later, that I had done well compared to so many of my patients to have reached the age of seventy, and promptly changed the subject.

Distracting myself from thoughts about my cancer and my future is just like my running. If I think about how much further I have to go, the running becomes exhausting, and I feel like giving up. If I manage to think about something else – a sentence I am working on, the next chapter in the FaceTime fairy stories for my granddaughters, or the design of a piece of furniture I am making – the miles pass much more quickly. But I have also learned to take a break in the middle of a run, and walk for a while, allowing myself to appreciate the scenery around me, instead of constantly pushing myself, trying to achieve something, making the present into an ordeal for a future reward. The lesson from my cancer is no different.

14

Six weeks after I had started on the hormone therapy, I drove down to a sawmill in Surrey, run by a retired colleague. It was the middle of January. A few days earlier it had snowed, but this had been followed by days of continuous rain, and I drove down the A3 in a shroud of rain, the windscreen wipers batting frantically.

My colleague had been a maxillofacial surgeon (and also a reserve officer in the Royal Navy) and in the past we had occasionally operated together on complex cranio-facial tumours. Now almost eighty years old, he had the pleasant but authoritative manner of a naval officer. It was a woodworker's nirvana – surrounded by stacks of enormous oak trunks, some still with a little snow on them, dissolving in the rain. I would look at these great dead trees and imagine all the perfect quarter-sawn boards that lay hidden inside them. There was a massive JCB for moving the trunks onto the twenty-foot bed of the mill. My colleague had recently undergone major surgery to one of his carotid arteries and had chronic back problems, but he continued to run the mill on his own. Over the years he had supplied me with some fine lengths of oak from which I had made a staircase, a table, and garden fencing. On this occasion I bought a cubic metre of sweet chestnut firewood, which just about fitted into the back of my car. When I got home, I found that my mobile phone was missing. I realised that it must have fallen out of my pocket as we loaded the firewood into the car. I had to drive all the way back again along the A3, in mist and pouring rain, cursing myself. Although I was fairly certain I would find my phone, I was absurdly panic-struck, as though my life depended on it. I had visions of my friend driving his JCB over it, but to my relief I found it unharmed, resting comfortably on a bed of wood shavings.

So when I finally got home, I was tired. I was annoyed to find an unmarked builder's van blocking the entrance to my garage, where I wanted to unload the firewood. A young man was sitting at the wheel, and I asked him irritably to get out of the way. Which he duly did, but I felt guilty about my bad manners and after parking my car I went over to the other side of the road where he had re-parked his van to apologise.

After I had done this, I walked back to my house. He got out of his van and came after me.

'Do you know you have a loose slate on your roof?' he asked, in a sympathetic tone of voice. He had a boyish, pink face, a strong Irish accent and a winning smile.

'Really?' I said. 'I can't see one.'

He managed to persuade me that one of the slates was slightly out of place. My eyesight is not what it was. I had been quite distressed a few nights earlier when my son pointed out the Orion constellation to me in the night sky over London, and I simply could not see it.

'I can fix it for you,' he said helpfully. 'We've been

doing some work for your neighbours,' pointing vaguely up the road as he said this.

'How much?' I asked.

'Oh, fifty quid.'

'Well, if you could fix it then, I'd be grateful,' I replied, thinking of all the problems I had been having with the recent awful weather and the various roofs I had built so inexpertly myself, and which leaked steadily when it rained. Maybe getting professional help would be a good idea.

His two mates suddenly appeared with a long ladder and one of them ascended to the roof. He returned holding a piece of wood which the first man showed to me.

'Look,' he said, 'it's quite rotten . . . I'm afraid many of the slates are loose and water is getting in. It needs to be fixed.'

'How much?' I asked.

He thought for a while, as though doing some calculations. '£1,400', he replied.

Looking back on this, I cannot really understand why I accepted this without question. I can only think that my recent diagnosis of cancer had made me vulnerable, and I was reaching out for sympathy and help. Perhaps the chemical castration for my prostate cancer was shrinking the critical parts of my brain, as well as my tumour, making me naïve and trusting. I accepted this preposterous offer.

'You will need to sign a contract,' he said, producing a pad of printed forms. 'Let's go indoors to sign it . . .'

But something in me made me refuse this, and instead we stood outside.

'Have you got a pen?' he asked, and I duly fetched one. I quickly signed an impressive-looking document and was shown an imposing Certificate of Guarantee in Gothic lettering.

'We usually ask for a deposit of 20 per cent,' he said.

'I haven't got £280 in cash,' I replied.

'Oh well, never mind,' he said kindly, presumably thinking that I was so stupid that I would soon be coughing up the £1,400 anyway.

He thanked me profusely for lending him a pen – a cheap ballpoint – which struck me as a little odd.

'You had better take the slates off,' he said to one of his mates, a tall man who looked wonderingly at me as he climbed up the ladder. I didn't think to ask the young Irishman why it was necessary to remove the slates in advance of putting up scaffolding the next day.

So he and his colleagues drove off, having left a hole in my roof, promising to return next morning.

I went back indoors and struggled to understand what had just happened. Two years earlier I had fallen for exactly the same scam, although with different actors. On that occasion I was feeling stressed and anxious over deeply unpleasant and horribly expensive litigation against a neighbour in Oxford that I feared would cost a fortune. This first con man had exactly the same play sheet as his successor – an offer to clear the gutters on the roof for a small sum was followed by the descent down the ladder with a piece of rotten wood, and the ominous declaration that the roof needed major repairs. Even the price quoted was the same – £1,400. The first scammers had also come into my house and enthused about how nice it was. They had even managed to put up scaffolding on the front of my house within hours of my signing a contract, before I realised that I had been a fool, and told them over the phone – far too politely – that I did not want their services. Perhaps there is a training school for roofing con men.

I sat down at the kitchen table and cursed myself for having been so pathetically gullible. I also disliked the thought of having to get out my heavy extension ladder and climb up onto the roof and inspect the damage. As a neurosurgeon I had seen far too many old men with terrible head or spinal injuries from falling off ladders. It was also difficult to accept that people could be so utterly venal and dishonest.

But eventually I dragged out my three-stage extending ladder and set it up against the roof. It was quite a struggle and I wondered if the chemical castration for my prostate cancer was starting to weaken me.

I found that the scamming roofers had removed half a dozen slates, breaking five of them in the process. There were two rents in the slating felt beneath the slates, which almost certainly the roofers had made deliberately as well. There was no rotten wood, and clearly there had been no water ingress. The slate which I had been persuaded was loose was several feet away from the slates they had removed and wasn't at all loose. I climbed down the ladder and rang the number on the contract I had signed. Somebody answered and shortly afterwards, somewhat to my surprise, the young man who had fooled me in the first place rang me back. He must have thought that I was so old and stupid that he could convince me that the roof needed repairing – which, of course, it now did, thanks to his colleague's efforts.

'I have been up and had a look,' I said, quite unable to overcome my deep need to be polite. 'There has been no water getting in.'

'Ah,' came the confident reply, 'you can only see the rafters from below. There's water getting in from above.'

'I don't think you quite understand,' I said, perhaps with a slight note of triumph. 'I have put a ladder up and looked at the roof myself. This is just a scam.'

'It's not a scam,' came the feeble reply, and that was the end of the conversation. Perhaps he had deceived himself into believing this himself, and that had helped him to deceive me, but I rather doubt it.

The basis of thc trick, of course, is that the victim cannot see what is happening on the roof and is entirely dependent on what he or she is told. So you pick a white-haired old man, who looks wealthy enough not to be too troubled about costs, flatter him to establish trust and then take him by the nose. But you probably don't expect the old fool to have second thoughts and a long ladder, and to climb up onto the roof himself after signing the contract.

The first time I fell for this, the cowboy roofers failed to remove the scaffolding that they had put up so quickly, despite my repeated telephone messages asking them to take it down. After many weeks I had climbed up the scaffolding and worked on the roof myself. There was, in fact, a problem, but not the one the scammers had claimed. The rampant wisteria that grows up the front of my home – I like the anarchic, overgrown look and the cascades of blue flowers in early summer – had forced its way between the gutter and the wall of the house, breaking the gutter and causing a leak. It took me a day to transpose the wisteria and mend the guttering with a steel plate and screws and sealant. This would certainly have been quite expensive if I had paid somebody to do the work. It was strange that the scamming roofers never removed the scaffolding. I discovered on eBay that it was worth several hundred pounds. Perhaps they had scammed a scaffolding company or perhaps they were worried that I would take legal action against them if they reappeared. I eventually took the scaffolding down myself with a spanner, swinging between the poles like an elderly gibbon. I rather enjoyed this. After waiting almost a year, I gave it all to a builder I know, who was delighted to have it.

As for the second occasion, my neighbours put me in touch with two local retired builders, Terry and Mick, who repaired the damaged work next day for a modest sum. As Mick supported the bottom of the ladder, while Terry climbed up to hammer back the roofing slates, he told me that Terry was seventy-nine, but he wouldn't tell me his own age. I took the opportunity of having the ladder up against the roof to prune the wisteria where it was growing over the slates. As I did this a passer-by stopped and spoke to me.

'May I just say how much I like the front of your house?' he said. 'I pass it many times and it always looks so welcoming.' I was touched by this unsolicited comment, in such contrast to the cynical, corrupting flattery of the cowboy roofers.

I told Kate about this sorry business.

'I think you did well,' she said kindly. 'How old are you? Seventy-one? Only conned twice in seventy-one years – not bad going. And isn't it better to go through life trusting people, even if it means being fooled occasionally?'

I also recounted the story a few days later to a friend of mine, an alarmingly clever but charming professor of economics, as we walked along the muddy Thames tow-path in Oxford, winding our way around enormous puddles.

'It's quite an interesting story,' I said, of my susceptibility to cowboy roofers. 'I can't really quite understand why I was so silly. I never thought that I could be so stupid.'

'That's what makes it an interesting story,' my friend replied.

But the story of my gullibility ends happily. I had saved a lot of money by repairing the roof myself after my first gullible episode, and after the second, I finally acceded to my family's demands that I stop doing all the repair work on my home myself. Instead, I employed Mick and Terry, who did the work much better than I could have done. And the litigation which had been so troubling me finally came to a conclusion after two years, four High Court hearings and 3,000 pages of evidence, mainly submitted by the neighbour. 'Full indemnity costs' were awarded against the neighbour – a very unusual outcome, lawyer friends told me – with a highly critical judgement against him from the judge.

The roofers had set out to trick me and did so by drawing me into what seemed to be a friendly conversation, and then with flattery duped me into trusting them. It is extraordinary how flattery – as any con man knows – is so effective in getting us to suspend critical thinking. In retrospect I felt as though I had been hypnotised. But it is also because once a conversation starts, it becomes increasingly difficult to withdraw, and instead the skilled con man takes control of you and leads you on. It is rather like misdiagnosis in medicine, as I would tell my trainees: the further you go down the wrong route, the more difficult it gets to go back and reconsider things. A colleague would probably be able to point out your mistake immediately, but your need for self-belief, and reluctance to admit that you might be wrong, can easily drive you on until it is too late. It is easier to carry on forward with blind optimism than to doubt yourself and go back self-critically. This is why good colleagues in medicine are so important, with whom you can discuss difficult cases and who will point out your mistakes to you.

When patients came to see me, I took their trust in me for granted. I did not have to try to deceive them in the way that the roofers had with me. Patients – at least in England – have little choice about which doctors they go to see, unless they pay to be seen privately. But even if you pay, and have made a choice of sorts, you still have to trust the doctor you see, for the simple reason that it is intolerable to think that he or she is incompetent or dishonest. In England, on the whole, patients trust doctors by default – and yet it was still very much in my interests to go out of my way to justify my patients' trust in me. Neurosurgery is dangerous and some of my patients were going to come to harm, despite my best efforts. I had to prepare them and their families for this – for 'complications' as surgeons call things going wrong – so that they continued to trust me despite everything. It is torture to treat patients if they no longer trust you.

In some of the countries I have worked in, the default position of most people is not to trust doctors. All countries have systems of licensing and regulation which are intended to enforce standards, but they are often bypassed in corrupt countries. One of the health ministers in Ukraine told me that she had made herself very unpopular by making it impossible for medical students to bribe their way through their final qualifying exams, which had been common practice. The students were now forced to work, and the teachers lost money. She did not last long in the job.

I did not have to trick patients into trusting me when

I first met them. But the problem with medicine is that it is uncertain – doctors deal in percentages, and like the weather forecast, can only be wrong if they make a prediction with 100 per cent certainty, which they never do.

We often hear the refrain 'The doctors gave me six months to live, and here I am six years later', when, almost certainly, the doctors had told the patient he or she *might* only live for another six months. As for the ones who did die within six months – we do not hear from them.

When I first met patients, the problem was not one of establishing trust – instead the problem was of preparing for the possibility of failure in the future. If treatment fails, how can trust be maintained? I would tell my trainees that the management of complications starts as soon as the patient and family come through the door of the outpatient room for the first time. This becomes easier as you become more senior – you are more experienced and confident, and all patients understand the importance of experience. But things will still sometimes go wrong, however experienced you are.

There are two ways of trying to maintain trust despite failure. The first is to show that you genuinely care for the patient, that you are interested in the patient as an individual, and that what matters to him or her matters to you. I suppose it is possible to fake this completely, but I rather doubt it. The second is that there are certain tricks – for want of a better word – that are important. You should always be seated when talking to patients, and never appear to be in a hurry. This is much more important than the banal formulae I believe some medical students are now taught, such as saying to the patient 'I would really like to get to know you better'. Patients are not fools and will quickly detect insincerity.

Patients respect doctors, on the whole, because their lives depend on them, but it only became clear to me after I had retired why most doctors really do deserve this respect. It is because of failure, and not because of success. There is nothing very special about success – it is wonderful, of course, for both doctor and patient. And yet if every operation and every treatment and every diagnosis was a success, there would be nothing very special about being a doctor. The triumphs are only triumphant, I would tell my trainees, because we have disasters. I remember on one of my trips to Ukraine I helped my colleagues operate successfully on a child with a benign brain tumour. We were breaking new ground by doing such dangerous surgery in that particular hospital and were jubilant. Shortly afterwards, we operated on another child with a tumour. Admittedly, a more difficult case and a malignant tumour. The operation seemed to go well. Next morning, before going into the hospital, my colleague and I went for a run in the local park, feeling extremely pleased with ourselves. It had been snowing overnight, and the park and city were looking very beautiful under a clear, winter sky. When we got into the hospital, we discovered that the child had suffered from a catastrophic post-operative haemorrhage, which had been missed overnight. It was my fault – I should have enquired about the post-operative care before the operation but had been lulled into a false sense of security by the previous case. The post-op care had been, in fact, very inadequate, and the child's critical deterioration from the haemorrhage not noticed until we turned up in the morning.

The child died slowly over the next few days. She was an only child, to a single mother. It was pure torture to talk to the mother each day, trying to maintain hope but gradually having to abandon it as her daughter's condition deteriorated. It is problems like this – living with failure, trying to maintain a trusting relationship with patients and their families – that makes medicine special. It is all too easy to hurry past the patient's bed, to avoid the family, to speak in half-truths or incomprehensible technicalities, and to deny both to yourself and to others that you could have done any differently. I did at least deliver an impassioned lecture to the hospital staff, after the child's death, about post-operative care, and can only hope it has made a difference. The pandemic brought my visits to Ukraine to an end, so I cannot know for certain.

PART THREE
Happily Ever After

15

I was on hormone therapy for six months before the radiotherapy could start. 'Fiducials' had first to be inserted into my prostate, which would be used as targets for the radiation. This involved visiting the radiation department in a different hospital from where I was to be treated – a hospital I used to visit once a week for a conference with my neuro-oncological colleagues about my brain-tumour patients. I went with my juniors on Friday mornings, and we would turn left on entering the hospital to go to the canteen, and have a large, cooked breakfast before attending the meeting. But now, once again, I had the slightly wry experience of entering the hospital as a member of the under-class of patients, no longer a self-important surgeon, and I turned right, to head for the radiotherapy department. I could feel myself lose height as I walked along the corridor.

The radiotherapy department was in the basement, and much to my surprise, rather beautiful (probably because it was not a PFI project). There was a tall atrium, flooded with light from large windows in a roof supported by slender columns. The atmosphere was calm and peaceful, akin to being in the cloisters of a religious building. It was a far cry from the crowded, often windowless outpatient waiting areas with which many of us are familiar. There was a helpful poster stating: 'Radiotherapy is cutting-edge. Radiotherapy can cure cancer.' I found this very cheering.

I was met by a friendly specialist nurse, and I was soon lying on my side in a small room, without trousers or pants but wearing one of those infuriating, disempowering hospital gowns that only does up from the back. At least on this occasion it made sense, as she proceeded to insert a proctoscope into my rectum and then, using ultrasound guidance, injected three gold fiducials into my prostate with a special syringe. It hurt only momentarily.

'How large are they?' I asked.

'Cylinders three millimetres long and one in diameter. Twenty-four-carat gold,' she said.

The price of gold has been going up steadily in recent years and I was most impressed.

'That was fun,' I said to her, as I hitched up my trousers.

'We've never had anybody say that before,' she replied, looking at me rather dubiously.

'Many years ago, I had to do one year of general surgery before training to be a neurosurgeon,' I explained. 'I had to do a rectal clinic on Friday afternoons. Neither the citizens of Sidcup nor I particularly enjoyed the experience, but now I have some idea of what it was like for them. It's always very interesting to be at the receiving end.'

I left clutching yet another printout, warning me of all the possible complications of the procedure. It was pouring with rain as I walked out to the car park. I thought of my three gold teeth, and how I could now modestly boast that both the entrance and the exit to my gastrointestinal tract were lined in gold.

The radiotherapy started two weeks later.

The radiotherapy treatment was at the Royal Marsden, back in in Chelsea, six miles from my home at the bottom of Wimbledon Hill. The quickest way to get to it was by bicycle – first along the Nature Trail beside the River Wandle and then along the Thames Path to the Albert Bridge.

My first session was on a Friday afternoon. It had rained heavily in the morning and the nettles and buddleias that line the path alongside the Wandle were bent low, so I had to take care to avoid them as I cycled past. The air was full of scent from the buddleias. The river arises, rather improbably, in Croydon. London once had many rivers – rejoicing in names such as the Effra, the Peck and the Quaggy. The Wandle is one of only two that have not been covered over and turned into sewers. Halfway along the path there is the surprising sight of a scruffy paddock with a high fence, and a few small ponies grazing behind it. I would often see children as I passed, pressed up against the fencing, face to face with the ponies. On the other side of the river are industrial sites, one a waste-collection yard with the constant sound of dustbin lorries disgorging their loads, an electricity substation with massive transformers and insulators, and then a patch of allotments. At Earlsfield I joined a major road for a short distance to cross the

Wandsworth one-way system, past the old Young's Ram Brewery – now turned into luxury apartments and renamed the Ram Quarter – and reached the Thames Path. There I joined other cyclists, runners, dog walkers and buggy-pushing mothers, to cycle past the dull phalanxes of new glass-sheathed apartment blocks that have come to dominate the banks of the Thames, and cross the river over the Albert Bridge.

Once I had arrived at the hospital, I was directed down to the basement. There was a small waiting area for the radiotherapy department, with a row of clerestory windows of frosted glass engraved with leaf patterns. I suppose it is considered inappropriate for the brave waiting cancer victims to be seen by passers-by, but as a result the patients and staff cannot see the outside world. There were only three other victims waiting for their treatment, all looking as well as myself, and looking neither victimised nor brave.

While I sat there waiting, to my surprise the oncologist appeared. He was wearing theatre scrubs.

'I didn't know you operated,' I said.

'Brachytherapy,' he replied. Brachytherapy involves inserting radioactive pellets into the prostate and is a different way of treating prostate cancer from the treatment I was to receive. 'Let's find a side room,' he said.

He took me down one of the many corridors, and we sat opposite each other in a small windowless cubicle. He asked about my latest PSA results, which he did not know, and we chatted for a while, mainly about his young son's views on climate change. It seemed that neither of us wanted to discuss my treatment or prognosis. I wonder whether he realised how much this simple human contact with him meant to me.

I had told myself fairly soon after my cancer was diagnosed that at the age of seventy it was absurd to hope or seek for a cure. All I should look for are a few more years of good life. Besides, even if I knew for certain that I had only months left to live, as opposed to years, I find it hard to see how I would live my life any differently. I have had a good and very fortunate life, I have a loving family, I have no bucket lists, and I do not want to rush around the world, although I miss Nepal and Ukraine and my friends there intensely. The trouble, of course, is that if I have a few more years, I will no doubt try to bargain for a few more, once my disease has returned. My urge to go on living is so overwhelming that it will only be overcome by intractable physical suffering, and even then I may well hope for just a few more days. But then again, I might not. I hope not.

All matter, in accordance with quantum mechanics, exhibits wave-like or particle-like behaviour, depending on the way in which it is being observed. Of the various types of radiation, medical radiotherapy mainly uses photons, the particles of electromagnetic radiation – the radiation which includes light. Photons are emitted by atoms when they are excited – meaning that incoming energy knocks their electrons temporarily into higher-energy orbits around the atom's nucleus. The electrons then sink back to their resting state and release energy in the form of photons as they do this. Radiotherapy uses much shorter, higher-energy wavelengths than the wavelengths of light, which are the wavelengths we can see with the rods and cones in our retinae. Our eyes have been shaped by evolution to respond to the particular wavelengths that have been useful to our survival and show only a fraction of what is going on around us. Insects see different wavelengths from the ones we can see, as to survive and reproduce they need to experience different aspects of the outside world from the ones we need – such as the ultraviolet iridescence of flowers, which bees can see. The bees in my back garden, with brains the size of pinheads, can range up to five miles away and yet return to within a couple of feet of their hive, by seeing the polarisation of light, to which we are blind.

Vision in all creatures depends on molecules called opsins, which change their shape in response to light, and trigger a nerve impulse. Different opsins respond to different wavelengths of light. The iridescent dragonflies, which in summer I can see flitting across the water of the pond I dug in my cottage garden, can have up to thirty opsins in the retinae of their enormous eyes. We have only three, which respond to certain wavelengths that the brain then interprets as being red, blue or yellow. Our brains combine these three primary colours to produce the other colours of the rainbow. A few women are thought to have four opsins.

Some researchers suggest that insects may even have

some kind of conscious experience. There are similarities in the structure of the midbrain – the upper part of the brainstem – between mammals and insects, and they suggest that conscious experience is generated there. 'Conscious experience' in this context simply means 'sentience', the ability to feel – pain, perhaps, or hunger – and not to think and be self-conscious in the way that we are.

You learn early on in your neurosurgical training that consciousness in your patients depends on both the brainstem and the cerebral cortex. You will occasionally see patients who come into hospital as emergencies with objects stuck through their skulls into their brains – a chisel, an electric plug, multiple nails from attempted suicide, or a piece of wooden fencing – I have seen all of these. There may even be brain tissue coming out of the wound. And yet the patient is wide awake, whereas a very small area of damage to the brainstem can render the patient deeply unconscious. So the standard model of consciousness is like one of those fibre-optic lamps that were fashionable forty years ago, with hundreds of strands, each glowing at the end. Consciousness is the brightness of the light produced by the hundreds of fibres. Turning off the lamp's switch – the brainstem – will produce darkness, but many fibres will have to be damaged to dim the light if the electricity supply is still intact. Small areas of focal damage to the cerebral cortex do not affect consciousness, but extensive damage will.

I doubt if we will ever know for certain if insects feel

anything. There is a similar problem with hydranencephalic babies – babies born with intact brainstems but minimal cerebral cortex. I was responsible for several such children and their devoted parents. They show many facial expressions – which look like joy, rage, distress – but there is no way of knowing whether these are simply reflex movements or if they involve some degree of sentience, of feeling. Some researchers assert that these babies have feelings. Their parents certainly think they do. There is a similar argument over abortion, and whether early foetuses can feel pain or not.

I cycled in each morning for the treatment and was rarely kept waiting for long. Once summoned, I would walk down the brightly lit corridor to the room with the great machine and lie down on the machine's couch.

'Henry Marsh. Five, three, nineteen-fifty,' I would rattle off like a private on parade and then pull down my trousers and pants. It was a matter of minutes to position me in the machine, a radiographer gently pushing the lower half of my body, made unattractively plump and hairless by months of hormone therapy, so that it was perfectly aligned with a laser light beamed down from above. The radiographers would then leave the room and I lay there on my own. I found it difficult not to attribute magical powers to the benign, giant machine and hoped that it would save me. It's a matter of quantum mechanics, something I know about but cannot even begin to understand as the mathematics involved are entirely beyond me. After a few minutes the machine would start slowly and thoughtfully to rotate around me. It made a strange sound as it moved – not unlike a distant chorus of mocking frogs, beside themselves with hysterical laughter.

It was a strange experience to lie in the great radiotherapy machine and know that my cancer (and everything else in its immediate vicinity in my pelvis) were being bombarded by destructive photons, which I could neither see, nor hear, nor smell – though I could certainly feel after-effects a few weeks later. And to know that my life – or at least a few more years of it – depended on these quantum-mechanical, magical, invisible rays.

There is a brand of profitable quackery which talks of 'quantum healing'. Quantum particles do things that are impossible in the everyday macroscopic world in which we live – they can be both waves and particles at once, become entangled and tunnel and be in two places at the same time. If impossible things like this can happen, it is claimed, then obviously other impossible things – such as being cured of terminal cancer – can happen. Large sums of money can be made telling fairy stories like this to the sick and anxious.

But there is nothing magical about therapeutic linear accelerators; they are triumphs of physical science and high-tech engineering, which have taken decades and thousands of scientists and engineers to create. A magnetron – just as in your microwave oven – generates radiofrequency waves (streams of photons). An electron gun (usually a heated tungsten filament) fires electrons down 'the wave-guide' – a long tube which lies within the radiofrequency waves. The electrons, steered by surrounding magnets, surf the waves, and are accelerated close to the speed of light to slam into a tungsten target. The target spits out photons as the high-speed electrons react with the tungsten atoms. The photons are then filtered and focused and shaped into a narrow beam, like the light from a torch, and fired into the cancer (and whatever else surrounds it). By firing the photon beam in a rotating arc, and modulating its intensity, as well as its shape, a high dose of X-rays is delivered to the tumour and a relatively low dose to the surrounding body.

Radiotherapy works by damaging all the cells in its path, in particular by damaging the DNA in the cells' nuclei. The high-energy photons knock electrons off the target's molecules, directly splitting the helical strands of DNA. This happens in both the tumour and in healthy cells, but the crucial difference is that cancer cells are less able than healthy cells to repair the damaged DNA and splice it back together again. Healthy cells are said to have 'redundancy' – they have several different mechanisms for DNA repair. There are also indirect effects where the radiation 'ionises' the water in cells. Water molecules – two hydrogen atoms bound to one oxygen atom by sharing two electrons – are electrically neutral, but if radiation knocks off a negatively charged electron, a positively charged molecule results: an ion. This process generates those notorious 'free radicals' beloved of quack medical treatments, in particular hydroxyl ions. These are pairs of single hydrogen and oxygen atoms, with a negative charge. They react with many other atoms and molecules, causing extensive damage to both cancerous and healthy cells. Recent research has shown that there are also 'bystander' effects, where cells outside the main radiation field also show changes. As with so much of science, the more you look, the more you find, and the more complicated it gets.

From a practical point of view, what all this means for me is that if my tumour dies, it will take months. The malignant cells with damaged DNA will – I hope! – die when they try to divide and grow. If my cancer had been contained within the prostate gland, so that the target for the radiation had clearly defined boundaries, I would almost certainly be cured. But I have probably left it too late, and my cancer has invaded locally around the prostate gland (in particular into the seminiferous tubules) and no longer provides a clear target. The chances are that there will be a few scattered cancer cells that the high-energy photons do not reach.

The treatment took one month – five days a week, with a total of 60 Gray – the unit in which radiation is measured when it is absorbed by matter, such as my body. Once I was positioned in the machine, motorised arms holding box-shaped structures came out on either side of me – it felt like a reassuring embrace – to take plain X-rays. These would show the gold fiducials buried in my prostate, which are used to aim the X-ray beam. The beam is targeted using a planning CT scan done before treatment. You have to have a full bladder and an empty rectum when this is done and the same applies to the treatment. This can cause problems. If the plain X-rays showed a satisfactory state of affairs in my lower intestines (meaning that they were not at risk from the radiation) a buzzer would sound. The massive gantry of the machine would then slowly and solemnly rotate all the way around me, irradiating me. As the plain X-rays were taken, the machine made a few businesslike clicks, but once the actual treatment started, I heard the chorus of frogs.

It all became very routine, and I quickly learned the sequence of the machine's moves and noises. Halfway through the month of treatment, however, there was a long pause after the initial X-rays had been taken. The radiographer came back into the room.

'I'm afraid,' she said apologetically, 'that there is too much' – and she hesitated momentarily – 'too much matter in your rectum. You will have to try to clear it out.'

'Oh shit,' I replied.

It took two hours of effort on my part and intense discomfort from my bursting bladder, before the radiation could be delivered. For the remainder of my treatment sessions, I would wait anxiously as the plain X-rays were taken and breathe a deep sigh of relief if I heard the buzzer sound, meaning that the matter-problem in my insides was satisfactory. I would smile happily as the machine started to rotate and irradiate me. By the time of the last treatment, the requirement to have a full bladder before the treatment was becoming very difficult, as the radiation causes severe irritation. I read up on the Internet about incontinence. I learned that there is a great underworld of incontinence and many devices available for dealing with it – ranging from nappies to penile clamps, via various catheter systems. Apparently, more nappies are now sold for adults than babies – further proof, if needed, of the profound demographic changes taking place in the modern world. I decided to carry a clean pair of trousers and pants in my bicycle panniers, as the threat of incontinence loomed ever larger. When I reached Battersea on my bicycle on the last day of radiotherapy I was quite desperate to urinate and dreading embarrassment and ignominy. Suddenly – as though by magic – a Portaloo materialised under the brick arches of Battersea Railway Bridge – like Doctor Who's Tardis, but without the accompanying electronic music. It must have been there all along, for the workers on a nearby building site, but I had not noticed it before. And it was unlocked. Uttering a prayer to the patron saint of toilets, I dived in.

The control of micturition, as doctors call urinating, is complex and poorly understood. There are some relatively simple reflexes between the stretch-receptors in the bladder wall and the spinal cord and brainstem, that result in reflex emptying of the bladder. But these reflexes are under the control of the cerebral cortex, where most neuroscientists think consciousness is to be found, and under conscious control. Frontal brain damage is often associated with incontinence. I have found that my irritable bladder, which increasingly dominates my life, becomes much less of a problem if I am busy or distracted, and my conscious attention is elsewhere. The waves of urinary urgency simply disappear, for instance, if the doorbell rings. There are various drugs which can lessen urinary urgency but most of them act on the brain, have 'cognitive effects' and are implicated in dementia. I was surprised that I was not told this when it was suggested I take them. Perhaps it was assumed that as a doctor I already knew this. In fact, I did not – but a little research on the Internet told me that I was better off without them.

The last treatment was not on the brand-new Bevan machine but on the older Brunel machine. There were sticky-tape marks on it, and it creaked a little as it rotated around me, in sympathy with my arthritic knees. There was no sound of mocking frogs.

16

With the onset of the first lockdown, I read – like many people – Defoe's *Diary of the Plague Year*. It was written sixty years after the Great Plague of 1665 and largely based, apparently, on his uncle's diary, but it reads with an extraordinary immediacy. In one passage he describes how people often died in agony. But it was spiritual agony as well as physical agony, as they feared they would go to everlasting hell if they had left their confession of sins too late. Defoe describes how, on their deathbeds, they begged others not yet afflicted to seek forgiveness of their sins.

Unlike Defoe's contemporaries and probably most of the opponents of Assisted Dying, I do not believe in any kind of afterlife. I do not believe I will be either punished or rewarded after I have died, or that I might become an unhappy ghost who haunts the world of the living. Admittedly, surveys show that most people in the modern world who believe in an afterlife believe only in heaven, and not in hell. Once you accept the message of neuroscience that thought and feeling are physical phenomena, an afterlife seems as unlikely as Boltzmann brains. There are a few neuroscientists I know who, somehow or other, manage to maintain a belief in an afterlife. One of them wrote to me, when he heard of my diagnosis, to tell me that our lives were just the canapés and pre-dinner drinks to the heavenly banquet, as he put it, that awaited those of us who had bought tickets. I am not buying the tickets. As far as I can tell, when I die, my constituent atoms, produced aeons ago in the supernovae of dying stars, will head off to be rearranged into other forms of matter and not to a heavenly banquet. We do, of course, live on, not as conscious entities, but as memories in the brains of those who outlive us. And so we have pyramids, chantries and headstones. And even those, of course, are ephemeral. And yet surely we have a duty to live our lives so that we leave good memories behind us, and not just funerary monuments and landfill rubbish?

I had struggled, at first, to come to terms with my diagnosis of advanced prostate cancer, but the more I thought about it, the clearer it became that the only question that matters is what my dying will be like.

As a doctor, I have seen many people die – some well, and some badly. There are many ways of dying. It can be fast, or it can be slow, it can be painless or painful, it can be horrible, even in the modern age (whatever some palliative-care doctors might claim to the contrary) or a peaceful fading away. And sometimes it is dragged out with intensive care and resuscitation, which all too easily can become a charade, a dance of denial. But only rarely is dying easy, and most of us now will end our lives in hospital (only a few of us die in hospices), in the care of strangers, with little dignity and no autonomy. Although scientific medicine has brought great and wonderful blessings, it has also brought a curse – dying, for many of us, has become a prolonged experience, even if severe pain is now only rarely a problem. Furthermore, modern diagnostic technology can predict our decline and death a long time in advance – just as it has done with me – while we are still relatively well and independent. Of course, we all know that death is inevitable, but everything changes once you have been diagnosed with a potentially fatal illness.

Prostate cancer typically spreads to the bones. This can be very painful and dying can be protracted. If the tumour spreads to the spine, paralysis of the limbs – of the legs, or of all four limbs – can occur quite a long time before death brings release. Surgery can delay the onset of paralysis if the spinal metastases have not yet caused severe compression of the spinal cord. The clinical picture, as doctors call it, is typical. The patient starts to develop back pain, which unlike most back pain, fails to improve after two to three weeks, and instead gets worse, and is often a problem at night. Sooner or later, the tumour starts to press on the spinal cord, either directly or by causing the vertebrae to collapse, and the patient starts noticing numbness and weakness of the legs. This then progresses, sometimes over days, sometimes over weeks, to incontinence and complete paralysis. Neurosurgeons describe the patients in this final stage as being 'sawn off'. If surgery is delayed until this state, it makes absolutely no difference. The patients remain completely paralysed until they die. If they are unlucky, this can take quite some time.

Operating on metastatic prostate cancer that has spread to the spine does not prolong life, but it is, as they say, better to die on your legs rather than off them. If you operate while the patient is still able to walk, it is usually worth doing, provided that the patient is likely to live for at least a further six months or so. I had cared for many old men with this problem. Until recently it was simple surgery – all you did was 'decompress' the spine, by removing the tumour and bone that was pressing on the spinal cord. But if the tumour had caused the vertebrae to collapse, surgery usually made things worse and was best avoided. The decision-making was easy. This was routine, simple surgery and the patients came and went quickly, being returned to the care of the oncologists as soon as possible. I would have no further contact with them. I knew that they would all die, sooner or later, and I thought no more about them. Since then, all manner of metal implants have been developed to deal with the problem of the collapsed vertebrae – complicated titanium scaffolding that straightens out the collapsing spine and keeps it rigid. It is now carried out by specialist surgeons – some orthopaedic, some neurosurgical. I stopped doing such operations many years before my retirement. The trouble is that the decision-making has become more difficult, and the balance of risk and benefit more difficult to assess. Should you carry out major surgery in advance of deterioration? At what point is the patient's prognosis too poor to justify major surgery? There is little point operating if the patient will only live for a few months. Complex algorithms with the long acronyms that doctors love (they make things sound more precise than is really the case) have been developed to guide decision-making.

But it remains a question of judgement and surgeons are fallible, and often too optimistic.

But there is nothing to be gained by operating on men who have already become sawn-off. The juniors are heavily criticised if they admit such cases, as false hopes have been raised, a precious surgical bed wasted, and all that can be done is to send the patient back to the referring hospital. And somebody has to break the bad news to the patient. I clearly remember one such case – a seventy-year-old man, completely paralysed from the upper waist down. This meant that not only was he incontinent and unable to walk, but also that he was unable to sit up in bed unaided. Having balled out my junior for admitting him, I felt obliged to talk to the poor patient. It was late in the evening, and he was lying in his hospital bed in a little ring of light from the bedside lamp, in one of those standard NHS four-bed bays, with curtains as a pretence for privacy. I drew the curtains and sat down beside him.

I introduced myself and he looked hopefully at me.

I cannot remember exactly what I said to him, but I remember telling him that there was no prospect of his ever regaining any independence. Given the severity of his paralysis, I could not even hold out much hope that he would ever get home, as he told me that his wife was disabled, and he cared for her at home. His children all lived far away.

He listened to me in silence. I wished him a good night and left. As I cycled home, I thought about what lay ahead of him.

The word euthanasia comes from the Greek and simply means a good death. The idea that doctors might help the dying to die is not a new idea. Sir Thomas More advocated it in his description of Utopia, but on the basis of a firm belief in the joys of heaven – the heavenly banquet – that would come to the devout after death. There are conflicting reports of the practice, or the absence of it, in our ancestral hunter-gatherer societies. The word came to acquire a deeply sinister meaning after its use to describe the mass murder of disabled and chronically ill people by Hitler's regime in the twentieth century – people who were dismissed as 'useless mouths'. Assisted dying is not euthanasia, as euthanasia means doctors killing patients without the patients' consent. Assisted dying, it cannot be over-emphasised, is about patient autonomy and choice. Assisted dying used to be called assisted suicide, which made it clearer that this was about a patient's decision and, of course, suicide is not illegal.

I have written in the past about my suicide kit – a few drugs legally obtained with which I should be able to end my life. After my diagnosis of advanced cancer, I became increasingly desperate as I imagined how miserable my death might be. I worried that my suicide kit might not work. I might, for instance, vomit up the overdose of tablets. In despair, I rang a close friend, a doctor who shares my views on assisted dying. I burst into tears as I told him of my diagnosis, and I could see Kate across from me, weeping as she listened. I discussed my cancer with my friend for a while.

'Isn't this a bit premature?' he asked.

'Yes. I know. But I want to prepare myself for the worst,' I told him. After a pause I said: 'Do you promise to help me when the end comes, if necessary?'

'I promise,' he said.

This promise helped take the edge off my agony, but did not remove it completely. I am very lucky that, as a doctor, I have the solace of access to an easier death than is permitted by the law in the UK, if this is my eventual wish. We are always free to commit suicide, of course, but it is difficult, as we have to do it violently, because access to suitable drugs is restricted. We have to jump from a height, or cut our throats, or kill ourselves by hanging or asphyxiation, or in countries such as the US with a gun, where there are some 20,000 such deaths every year. These unassisted methods are distressing – both for ourselves and for our families. An undertaker I know once told me how she often came across such deaths in elderly people, sometimes from deliberate self-starvation, and how they often were not classified as suicide. She told me of the misery and guilt they caused in the families of the deceased. Alternatively, you can go to one of the clinics in Switzerland, such as Dignitas in Zurich, if you have the money – it's not cheap. But with all due respect to the Swiss, I do not want to die in Switzerland. Like most people, I want to die at home.

Of course, I fully accept that it might not be my wish

to commit suicide when the end comes. Many of us prefer to let nature take its course, although often by slipping into a dissociative state, where part of our mind knows that we are dying, and part of us thinks that we will go on living. I had observed this in my mother as she lay dying, and in several of my own patients. I had also started to observe this in myself – not that I was yet dying – as I fluctuated wildly between hope and despair when I first learned of my diagnosis. It is impossible to feel opposite feelings simultaneously, and instead we alternate between them, rather as we do with optical illusions. At the end, faced by the certainty we no longer have a future, but unable to abandon hope, the coherent, unitary self, which neuroscience cannot even begin to explain, breaks down into separate parts, with dissonant beliefs.

In the wake of the mass murder of patients and concentration-camp victims by Nazi doctors and also of medical experiments on unwitting patients by doctors in many countries, the Nuremberg Code was promulgated in 1947. This in turn led to the development of the so-called Four Pillars of medical ethics – autonomy, beneficence, justice and non-maleficence. Patient autonomy is central. Patient autonomy means we are legally entitled in the UK as patients to refuse further medical treatment, even if this will result in our death. In other words, we are allowed to choose to die, but not how, when or where. Only the doctors can decide that. And yet, as the great reforming judge Lord Denning once observed, the illegality of assisted dying in Britain means that it is illegal to help somebody do something that is not illegal. Surely this is wildly illogical?

Many countries have now legalised assisted dying – for instance, Belgium, Canada, Spain, New Zealand, Germany, several American states, Austria and the Netherlands. The list steadily grows. Different countries have different criteria that need to be met, and there are different ways in which it can be carried out. In the Netherlands it can be by lethal injection, in California the patient has to drink a cocktail of drugs (leading to all sorts of absurdities to enable paralysed patients to receive it). It can be granted on the grounds of intractable suffering, or only if the patient has a terminal prognosis of less than six months. In most of these jurisdictions, independent experts must review the request and there must be a delay of some weeks before the request is granted. Nowhere is there assisted dying on demand. In the Netherlands, there has recently been a campaign to allow assisted dying on the grounds that you have lived a 'complete' life, without the need for a terminal diagnosis or intractable suffering, but the campaign so far has not been successful.

In Britain, an influential minority – in particular some senior palliative-care doctors, some disability rights activists and certain MPs – is implacably opposed to assisted dying in any form. I suspect that most of them have religious faith but their faith, of course, is irrelevant to the debate – to refer to it would be argument ad hominem: criticising a person's argument on the grounds of his or her motivation for arguing it, which has no bearing on the validity of the argument itself. Politicians charged with corruption invariably invoke it, saying that the case against them is 'politically motivated' – but that has nothing to do with whether they acted corruptly or not.

A bill for the legalisation of assisted dying in the UK was defeated in the House of Commons in 2015, even though repeated surveys of public opinion have shown that an 80 per cent majority favour the legalisation of assisted dying. The Hansard record of their debate makes painful reading in places, such as the claim by one MP that the drugs used for assisted dying cause the patient to choke to death and are 'painful and barbaric' – either an outright lie or the product of staggering ignorance. Another MP in the debate used a frequently cited argument against assisted dying. He said that it was unacceptable because an opinion poll in Oregon, where assisted dying is permitted, showed that 50 per cent of people requesting it said they did not want to be a burden to their families. He failed to point out that the same poll showed a far greater number were more concerned about the loss of autonomy that dying so often involves. I find this 'burden' argument rather strange. I love my family and therefore do not want to be a burden to them. They love me in return and probably would want to care for me more than I want them to, but nor do they want me to suffer, and they do not want to be left with sad and painful memories of a drawn-out and wretched death.

But this is something for us to sort out ourselves, subject to certain legal safeguards – we do not need sanctimonious MPs and God-fearing doctors telling us how to live and die. Besides, my own anecdotal experience is that so often it is the patient who is ready to die but the family who cannot let go. It is they, after all, who will be left behind. Death is not just about the end of an individual life, but also the grieving lives of those left behind.

The availability of assisted dying in many countries means that we have evidence now as to how it works in practice. Until recently there was no evidence and therefore no way of rejecting the various hypothetical arguments against it. But this is no longer the case and none of the objections from its opponents in this country are being realised.

The opponents of assisted dying cannot argue against it on the grounds of patient autonomy, as suicide is not illegal, so they have to introduce a new idea: that there are many doctors, relatives and healthcare workers who will want to persuade or pressurise vulnerable people to ask for assistance in killing themselves. The dying, the disabled, the elderly, we are told, are 'vulnerable'. Society will become corrupted by the legalisation of assisted dying and the lives of these vulnerable people as a whole devalued. The opponents cannot produce any evidence to support this argument, which I find implausible in the first place. It is an entirely hypothetical argument and is not borne out by what is happening in countries where Assisted Dying is permitted. It simply passes over the existence of legally enforced safeguards. The argument is very similar to those used over the years by the tobacco lobby against banning smoking and the fossil-fuel interests against measures to mitigate the coming catastrophe of climate change. The aim is to sow doubt – and is an argument to be used when you either have no evidence in support of your position, or the evidence is against you.

Elderly parents or grandparents, it is claimed, will be bullied into committing suicide by their families – to get rid of them, or to get hold of the family silver – either explicitly or by abusing or neglecting them to a point where they find death preferable. In other words, if suicide becomes too easy and insufficiently unpleasant, these vulnerable people will be made to feel such a burden, or their lives made so miserable, that they will want to die. There is a glib little sound bite for this – 'the right to die will become the duty to die'. But those who say this fail to admit its corollary that no right to die imposes a duty to suffer. There is no evidence whatsoever that any of this is happening in countries where assisted dying is allowed. Where are the queues of elderly people asking for assisted dying for the greater good of society? Or do we really think British families and healthcare staff are especially callous?

It seems a strange idea that banning assisted dying is a dam holding back callous societies and a flood of murderous carers. Abuse of the elderly certainly occurs in our society, but the safeguards with which assisted dying comes will do more to identify and prevent such abuse than the present arrangements, or rather the lack of them. There are various forms the safeguards can take, but in principle independent experts must establish that the person requesting assisted dying has mental capacity, is not clinically depressed, is aware of the alternatives, and is not being coerced. Will the old and frail be so terrified of their persecutors that an independent expert will not be able to see that they are being coerced into killing themselves?

A different argument is that the legalisation of assisted dying will be used as an excuse to provide less hospice care. Again, this has not been borne out in countries where assisted dying is available. Assisted dying should be seen as *part* of palliative care and not in opposition to it, as is the case in countries where it is legal. In England, it is perfectly legal for doctors to prescribe very large doses of opiates to dying patients, even though it is possible that this will hasten death by causing respiratory depression. The *intent*, it is argued, is not to kill but only to ease suffering, and hence the Suicide Act, forbidding assisted dying, does not apply. I am assured by palliative care doctors I know that this 'terminal sedation' is rarely used, and if used it is only after discussion with the patent and family. This was not the case when I was a young doctor many years ago. But if patients are allowed to choose painkilling treatment that might hasten their death, and which will certainly end with their death, why can't they choose treatment (i.e. assisted dying) that brings their death swiftly, rather than slowly?

Opponents point to the Netherlands and say that permitting assisted dying has been a slippery slope. They say that too many people are being allowed to die prematurely. The number of assisted deaths has increased slightly in recent years. Opponents say that this is terrible, and an inevitable consequence of legalising assisted dying in any form. Whether the situation in the Netherlands is terrible or not is a matter of opinion, but there has not been a similar rise in all the other jurisdictions where assisted dying has been legalised. It is all a question of how you design your legal safeguards and of national culture. The Dutch are famous for plain-speaking, and do not shy away from difficult conversations in the way that the English do.

I know many people of my age who have had to watch at least one of their parents descend into the dementia that so often comes with advanced age. My own father died at ninety-six after more than ten years of sad decline. Most of us fear the same for ourselves but feel helpless in the face of it. Assisted dying is of little help, as in all countries where it is legal, the person asking for it must have mental capacity – i.e., not be demented. It is sometimes possible to diagnose some of the dementias in their early stages while you still have mental capacity, and some of the British people going to Dignitas have been in this situation. This calls for a resolve and determination that is probably beyond most of us. It is also true that in the Netherlands you can write an advance directive saying that if you lose mental capacity from dementia, you wish to be considered for assisted dying. But few doctors in the Netherlands are willing to carry this out – and I certainly would not feel able to administer lethal injections to demented old people, even though it is what I want for myself were I to survive cancer and become demented. But my demented self might be a happy vegetable (although it may not) and not at all keen to die. I see no easy answer to this problem. Assisted dying – despite what its opponents claim – has little relevance to the problem of the ever-increasing number of people with dementia.

It has always struck me as somewhat illogical that the most passionate opponents of abortion and assisted dying usually have religious faith, with a concomitant belief in life after death. Surely, if our lives continue after death, abortion and assisted dying are not absolute evils? If there really is going to be a heavenly banquet after death, why delay? It is as though they think that assisted dying is cheating – that we need to suffer when dying if our soul is to be reborn, that there is something 'natural' about dying slowly and painfully. Some even claim that dying is a transcendent experience, as Tolstoy describes in his novella *The Death of Ivan Ilyich*. I find this distasteful. If there is any transcendence, it is more likely to be enjoyed by the witnesses, and not by the dying. The transcendent experiences reported by some people who have had 'near-death experiences' – such as being resuscitated after drowning or a heart attack, or falling from a great height onto snow-covered trees – are probably peculiar to being faced suddenly by the certainty of death with full awareness and are quite different from protracted dying in bed.

Our fear of death makes it very difficult to look it in

the face and see the manner of dying as a practical problem, as a question of choice, that can be regulated by the law, rather than as something divinely ordained, and which is not negotiable. We all fear death, but for people with religious faith there is the added fear that their faith might be mistaken, that there is no human soul or essence and no afterlife – that death might be final, with nothing to follow. That we are our brains, which like our bodies are made of matter, of atoms and elementary particles, the ashes and dust of stars.

To help somebody to a peaceful and dignified death that they have chosen for themselves is an act of care and love. The opponents of assisted dying insist that if my death involves great suffering, it is my duty to endure it to the bitter end. They have produced no good evidence to show that my suffering would be of benefit to anyone else. They are also quite unable to see that there is a profound difference between assisting and encouraging somebody to end their life. They claim to be compassionate but in reality are responsible for much suffering.

17

After my radiotherapy finished, I started to attend meetings again at the neurosurgical department where I had previously been the senior surgeon. At first I found it very difficult, but I felt myself increasingly drawn back to being a doctor – not to treat patients, but to teach. The Covid pandemic had brought my work overseas – mainly in Nepal, Ukraine and Albania – to an end, and I was becoming a little bored, as well as finding it difficult to get out of bed in the mornings. Like most surgeons, I have always seen teaching the next generation of surgeons as an integral part of surgery. Surgery is a practical craft, and the relationship between master and apprentice can be deeply fulfilling. I missed it, whereas I did not miss the operating at all, even though I had loved it so much at the time. Like many retired surgeons I know, I take more pride in the successful careers of surgeons I have helped train than in all the patients I treated over the years. It is a great privilege to be part of a great tradition. Everything you do as a surgeon, even though it feels so intensely personal when you operate, is the culmination of the work of countless surgeons who went before you. You stand on the peak of a mountain of pebbles and, if you are lucky, you might add a few pebbles of your own.

Many years ago, I had introduced a meeting every morning in the department where the senior and junior doctors came together to discuss current cases. This was explicitly not a 'handover' meeting, where patient details are handed on from doctor to doctor, but a meeting to discuss how patients should be diagnosed and treated, and the decision-making involved. The meeting is about teaching, and team-building, and not just a dull recitation of facts and figures. The on-call junior surgeon presents the case, and I then question the audience of junior doctors about the diagnosis and how the patient should be managed. Doctors talk of the patient's presenting history, and it is indeed all about storytelling. I loved these sessions.

It is a creative act to construct a diagnosis from the story of the patient's symptoms and the 'signs' found on physical examination. And after constructing a story we look at the scans and see if we have been correct or not. I am often wrong but it's important to demonstrate just how fallible I can be. I teach now as a patient myself, with cancer, and I try to convey to the trainees just how great is the distance between themselves and their patients, and how they must try not to be corrupted by the power they have over their patients. I don't remember receiving any teaching about how to talk to patients when I was being trained, but at least undergraduate and postgraduate medical training now does include some training in communication.

After looking at a brain scan showing some dire problem, with the patient unconscious and *in extremis*,

I will often ask: 'What will you say to the patient's family?'

'I will say that the prognosis is poor . . . er . . . that the intracranial pressure is high and . . . er . . . the brain shows shift . . .' the trainee will start to say, but I usually interrupt.

'Oh come on!' I will shout. 'What the fuck does that mean? The family won't understand any of that.' And then we will discuss how to break bad news in clear and simple language. 'Always sit down,' I will say. 'And never appear to be in a hurry, even if you are. And keep it simple. Say as little as possible. Don't be frightened of silence.'

There is an art to conducting meetings like this. I would like to think that I do it well, although I have to struggle not to talk too much and dominate the meeting. I must make an effort to single out individuals by name to answer the questions – but this is made difficult by the fact that there are now thirty junior doctors in the department, and I can never remember their names, let alone recognise them behind pandemic face masks. If you direct questions to nobody in particular you are usually met by silence. And the threat of suddenly being asked a question by me, and then mocked if you haven't been listening, forces the juniors to pay attention. It is important to make jokes and tell stories, especially about the mistakes I have made in the past. As a senior doctor, one of the most important things you should do is to try to help your juniors avoid making the mistakes you have made yourself.

But going back to these morning meetings has come at a price. I have lost much of the godlike detachment that I had in the past. Quite often the case of an old man with progressive paralysis from prostate cancer will be presented. I feel a stab of sickening dread and anxiety, knowing that sooner or later there will be a clinical meeting where it is my end-story that is being told, with my name on the scans.

I had bought the derelict lock-keeper's cottage on the Oxford Canal shortly before retiring, as I planned on leaving London and living in Oxford, where Kate lives. I had grown up in Oxford until I was ten years old. Kate has had an apartment in the city centre for many years, but there is no room in it for all my tools and books, and I planned on putting these in the cottage. But Covid and cancer changed all this. My three children and three granddaughters all live in London, and I realised – as do most people towards the end of their lives – that I wanted to be near them. When my children were young, I had always put my work first and neglected them. I do not want to repeat this mistake with my grandchildren. But this change of mind meant that I have become increasingly uncomfortable about having two homes that are empty much of the time, just as I have decided that the days of my flying around the world to teach and lecture need to come to an end. So rather than sell my house in London, I have decided to sell the cottage, despite the seven years of hard work I have put into renovating it.

Having made this decision, I am surprised to find that whenever I go to the cottage – usually by bicycle along the towpath, given the lack of road access – I do not feel especially sad about relinquishing it. Instead, I feel a deep satisfaction that I rescued it and gave it a new life. Other people will now live there, long after my death, and this seems much more important than any use I might get from it during what is left of my life.

One of the first things I did after buying the cottage was to plant six apple trees in the garden – Cox's orange pippins – and a walnut tree, *Juglans regia.* I also planted hundreds of daffodil and tulip bulbs. I built a swing for my granddaughters – bright red, and styled, after a fashion, on a Japanese *torii* gate. It faces the neighbouring lake, with its reeds and swans, and the railway embankment in the distance. There is usually a heron, standing as still as a statue in the shallows, but as soon as it sees me it spreads its wings and flies slowly away.

There were two battered sheds in the garden built of rusty black corrugated iron – a vernacular style you rarely see now, and which I like. I employed builders to dismantle one of them and rebuild it on a solid floor, using the original corrugated iron with a few extra sheets. I put wooden trellis on the walls and planted climbing roses, which I hope will flourish. This was to have been my new workshop, replacing the one I have in London.

On a recent visit, after struggling to hang the workshop's very heavy, newly made double doors, I cycled back to Kate's flat along the towpath. There was a helicopter hovering like an angry hornet over Oxford's city centre.

'There's a "Rhodes Must Fall" demo in the High,' Kate told me, when I got back. Cecil Rhodes – the racist British imperialist at the turn of the nineteenth century – had given a lot of money to Oriel, one of the Oxford colleges, where he had once been a student. There is a small statue of him high above the High Street, beneath a roof gable of one of Oriel's buildings. Until now, I had never noticed it.

'Attacking statues is silly,' I said.

We argued about this for a while. Kate was all for action of some kind against the statues of slave owners and racists. I responded that it was a slippery slope, with no clear end in sight. Removing statues could not undo the past, I said. Where do we draw the line? We draw it somewhere, she replied, and we can debate where that should be.

I went to bed feeling rather angry – Kate and I rarely argue. When we do, she is almost always in the right. It can be very annoying.

Since being diagnosed with cancer, I have become unable to sleep through the night. I do not know whether this is an effect of the hormone therapy, or the restlessness and unhappiness that comes with having a potentially fatal illness. Perhaps it is just old age. I get to sleep easily enough but invariably find myself wide awake in the middle of the night for an hour or so. I have learned to accept this, and I pass the time constructing fairy stories for my granddaughters – this usually gets me back to sleep. I suppose what I am really doing is telling myself a bedtime story. After the argument with Kate, I woke in the early hours from a terrifying nightmare. I dreamed that I was in the house in London where my family had lived after we moved from Oxford.

It was a Queen Anne terrace house in Clapham. Clapham was a very unfashionable area in 1960 when we moved there from Oxford – which was why my father could afford it. It had been turned into bedsits, and there were at least six bathrooms that had to be stripped out before we could move in.

It was a very beautiful house, three storeys with a basement, the rooms with high ceilings and tall sash windows looking out over Clapham Common. The rooms were all wood-panelled, the floors with wide floorboards, and each room with a fireplace with a fine cast-iron fire-basket. There was a grand oak staircase, with barley-twist bannisters. Even now, twenty years after we moved our demented father out of his grand home into a nearby flat and sold the house, I can walk in my mind's eye through each room and remember every detail. The creak of certain floorboards, the feel of the staircase's carved handrail in my hand, the smell of books in my father's study.

In my dream I was in my own room at the top of the house. I had lit a fire in the fireplace. And then I realised that the fire had spread, and the house was starting to burn down. I looked down through a crack in the floorboards and could see below a raging inferno. It was like looking into a volcano, with magma churning in the depths. I desperately poured a small jug of water onto the blaze, but the flames became fiercer and fiercer. I rarely have nightmares, but when I do, I usually realise I am dreaming and wake myself up. So I woke myself up, but the deep anxiety of this dream stayed with me well into the morning.

It seemed to me that the dream expressed a deep fear – for some people mixed with guilt – which afflicts many privileged, affluent people, of people less fortunate than themselves. I had refused consciously to accept Kate's argument, but it was as though my unconscious, sleeping self had accepted it and turned this acceptance into the story of the dream. As I thought about the dream, I realised that I had undergone a complete sea-change in my opinion and my feelings about the problem.

I had been taught about the slave trade as a child, and I knew of its importance in financing Britain's Industrial Revolution and the war with Napoleonic France. I was taught this as something which had happened a long time and was of no relevance to the modern world. But when I think of all the descendants of Britain's slaves and of other subjugated races who live in Britain, and who are as British as myself, I see things very differently. Why should they have to look at statues glorifying their ancestors' oppressors, without there being any kind of admission of guilt? If we take pride in our nation's history – as the current government demands that we do – should we not be ashamed as well?

And then there is the sight of desperate immigrants crossing the Channel, or freezing on the Polish border, escaping from countries ruined by war and civil war, where the weapons used are almost all manufactured by the 'advanced' nations, and where the wars are so often about the mineral resources our so-called civilisation requires. I can see less and less of which to be proud.

And as climate change takes its toll, and sea levels rise, the waves of immigration now will be nothing compared to the ones to come. My granddaughters will live in a profoundly different world from mine.

18

Covid lockdowns brought Zoom and FaceTime: storytelling on my iPhone for my three granddaughters. I did this almost every evening for two years and continue to do so. Sometimes, a little desperately, I have to invent a story even as I tell it.

It started easily enough, although I am a little embarrassed by how often the stories were derivative. But then it has been said that there are a limited number of themes in human storytelling, centred on the so-called monomyth of a hero or heroine and a quest, with transformation and return. In my stories the leading character was – of course – a girl, although so many subsidiary characters appeared as the months passed that I had difficulties remembering their names. To my relief, my granddaughters usually did as well.

The principal character was called Olesya, a Ukrainian name, in recognition of my love for that troubled and marginal country. Olesya lives in a house in England with her Ukrainian aunt. The backstory of how this came about and what has happened to her parents remains obscure. Parents are an inconvenience in most children's stories and need to be got out of the way as soon as possible. In Olesya's bedroom there is a magic door that can only be opened when there is a full moon at midnight. The sky must be clear of cloud, so that the moon lights a pathway across the oak floorboards of the bedroom to the magic door. The door probably originated in the Carpathian Mountains. I'm not quite sure if it can be opened when the weather is cloudy, but fortunately this never seems to happen.

The door opens to a field of flowers in Fairy Land, with a path leading to Fairy Castle.

At first the stories were simple and obvious enough – an evil rain witch, for instance, who makes it rain all the time in Fairy Land, causing floods that threaten Fairy Castle. Olesya sets out on a quest to confront the witch, and meets talkative, magical animals on the way, who give her magical weapons with which to defeat the rain witch. Olesya does, however, also use the cantilever principle to build a bridge using tree trunks and ropes to cross a ravine, at the bottom of which are ravenous crocodiles. The crocodiles have a passion for bananas, and these can be used to divert their attention when building the bridge. Fortunately, there are many banana trees growing nearby. The rain witch lives in a three-dimensional maze, an ascending spiral, at the heart of which is a doll's house. The witch lives in this, having miniaturised herself. A magical tool for seeing through walls, donated by a helpful serpent, proves invaluable in negotiating a way through the maze. The rain witch has a bathtub for a bed, with the shower above it turned permanently on. Olesya manages to miniaturise herself with the help of a magical tiger and defeats the rain witch by pulling the plug in her bathtub. She then turns her into smelly green dust with a battery-powered hairdrier, a present from another magical animal. It subsequently transpires that the rain witch has an evil wizard brother who revenges himself on Fairy Land with a large smoke machine that blocks out the light of the sun. Olesya defeats him with the help of some friendly local dragons, led by their chief, the red dragon Razubel.

There is a large library in the attic of Fairy Castle, which has many useful volumes. Olesya's friend Christabel lives with an orphaned unicorn who develops the dreaded Droopy Horn Disease. The unicorn – called Florinda, as far as I can remember – is so embarrassed by this, that she refuses to be seen in public, and hides in her stable. Fortunately, Olesya finds a book entitled *Unicorns and their Diseases*, and Florinda is cured, but I have forgotten how this was done. In a later story, my granddaughters learned that Florinda had become orphaned when the ice floe she was standing on with her parents broke up as a result of global warming.

In one story, Razubel becomes deeply depressed and refuses to come out of his room in Dragon Castle. It turns out he is hopelessly in love with the beautiful white dragon Eddfa, but her family are terrible racists, and will not hear of Eddfa marrying a red dragon. The white dragons live beneath a glacier and are subject to vicious attacks by ice monsters. When Olesya and Razubel overcome the ice monsters, the white dragons lose their prejudices, and Razubel and Eddfa can marry.

If there are only a limited number of stories, the number of directions in which you can travel is also limited. You can cross oceans and deserts and climb up mountains (all of which Olesya did on many occasions) but you can also go above and below them. So, there was a story about a sea monster who was kidnapping mermaids and chaining them up as living statues in his garden at the bottom of the ocean. Olesya rescued them, and overcame the monster, by using a submarine, built by Inginia the Engineer Fairy, who played an increasingly important role in the stories. Inginia was able to make almost anything. This included a tunnelling machine, so that when sinkholes appeared in Fairy Land, and Fairy Castle started subsiding, Olesya was able to visit the earth elves, who were causing the problem with too much digging. She showed them how to shore up their tunnels with pit props. She was rewarded with emeralds and rubies by the king of the earth elves. While underground, she discovered a race of dragons who had become trapped there many aeons ago, and who had lost the ability to fly. Using her tunnelling machine, she led them back to the surface of the earth, and her friend Razubel gave them flying lessons, as they had forgotten how to fly while living underground. Olesya gave Razubel the emeralds and rubies, which he wore on a chain around his neck.

But I missed an opportunity here – I should have made the dragons flightless like dodos and introduced my granddaughters to Darwinian principles of natural selection. (Of course, their vestigial wings could always have been quickly restored, and evolution reversed with some handy magic, once they were up in the open air again.) Olesya and Christabel, riding on dragons, were able to meet the benign, grey-eyed cloud creatures who live in the sky. The cloud creatures were very grateful to the girls for having rescued a young cloud creature who had sunk to the ground. They had used a pump, built by Inginia, to re-inflate him and get him back up in the sky again. Olesya had also served as an arbitrator when there was a major row between the cirrus and cumulus cloud creatures as to who should be highest in the sky. But imagination flags, and by 7 p.m., with bedtime approaching, my granddaughters were disappointed to learn that cloud creatures have rather limited conversational abilities and had little to say to the girls, and the story came to an end.

Inginia also made a space rocket so Olesya could visit the moon. The moon in Fairy Land is quite unlike the moon in Olesya's world. There are moon fairies and butterflies with long, furry tails – the end result of a failed attempt by Olesya to use biological means to control an infestation of mice on the moon. The details are complicated and need not detain us but involved cheese. She built a giant space telescope on the dark side of the moon (which, of course, is not really dark) and there she met the Moon Bear Fondrok, who has fur of pure silver, and with whom she explored the universe using the telescope. Fondrok had a great fondness for chocolate, which resulted in addiction. But Olesya helped him kick the habit.

Olesya was not transformed by her adventures, in contradiction of the monomyth. Instead, the stories became transformed – at least magic became muddled with engineering. Inginia, the Engineer Fairy, achieved near-magical results in her workshop. The king and queen of Fairy Land singularly failed to provide leadership when Fairy Land was invaded by aliens from outer space, and the situation was only saved through Olesya's efforts. There was then a revolution and Inginia was democratically elected as leader of the fairies, but my granddaughters nevertheless insisted on referring to Inginia as the queen of the fairies, despite my objections.

My granddaughters particularly liked the stories about the seven baby dragons, the offspring of Razubel and Eddfa, who were hatched from eggs. There were two red babies, two white, one white with red polka dots and one red with white polka dots, each a male and female pair. The seventh was pink, and of uncertain gender, but much loved by all the other dragons.

Olesya and Christabel had to rescue the eggs, and the babies after they were hatched, on more than one occasion from evil witches, who were crazy for designer handbags made from baby-dragon skin. The witches were evil, but young and beautiful, and terribly vain. They were not toothless old women with facial hair and crooked noses. Although both the girls had magic swords, I was a little reluctant to have too much violence in the stories, even though extreme violence is such an important part of so many of the most famous children's stories. I therefore introduced the rule of law into Fairy Land. If you made a promise, you would drop dead if you broke it. By this means, Olesya and Christabel righted many wrongs without beheadings (although the occasional beheading of particularly obnoxious goblins did slip in).

Should fairy stories be an escape from the present or a preparation for the future? They are a form of play and play in children and young animals is universal. How can you doubt that animals have feelings just like ours, when you see young animals playing together? Evolution has made the young playful, as our brains cannot develop without play. It is a form of preparation for independence and yet at the same time a celebration of the wonder of the world and our ability to imagine worlds that do not exist.

Neuroscience tells us that reality is a construct built by our brains from only those aspects of the outside world that we have needed to perceive in order to survive and reproduce. We live in a model of the world, a story of sorts. The intense feeling of plot and narrative in our dreams, although it may well be meaningless, suggests that making sense of the world by telling stories is a critical part of being human. Fairy stories are about what doesn't really happen, which is why typically they end: *and they all lived happily ever after.*

19

A year has passed since I was diagnosed with cancer. I have now joined the great underclass of patients with treatable but probably incurable disease, whose lives are ruled over by doctors. Our lives lurch from scan to scan, in suspended animation, from blood test to blood test. And yet, given my age, nothing very much has changed. I am approaching the end of my life in any case, and to be cured only means to die from something else.

I have not found it easy to come to terms with the proximity of my own death – either from the cancer for which I am being treated or, if the cancer treatment is successful and I am cured, from the dementia that I fear even more than death. Of the two possibilities, dying from cancer is surely preferable. This is not an easy or comforting line of thought. If I have to die from cancer, and if the dying is going to be distressing, I hope that assisted dying will be legalised in this country in time for me, and I have some choice in the matter of how, when and where I die. But I still find it hard to escape my deep biological optimism – with which evolution has both blessed and cursed me – that all will be well, that somehow I will be saved and death avoided. I can see the same folly in my initial insouciance – shared by so many others – about the Covid pandemic. Worst of all, I see it at work in the unfolding catastrophe of climate change – our clear understanding of what will happen if no action is taken and our failure to take effective action. Like most people my age, I am haunted by the thought of what kind of future awaits my granddaughters, and the ruined planet my generation might well bequeath them. But we have a duty to be optimistic – if we are not, and therefore do nothing, then the worst will certainly happen.

At nine o'clock in the evening, as it was getting dark, the stars just starting to appear, I walked along the towpath from my lock-keeper's cottage, pulling behind me the small and battered four-wheeled cart I use for carrying building materials. On this occasion, however, I had loaded it with cushions and blankets. I met my daughter and her family at the lift bridge, where the track to the canal ends. My granddaughters happily climbed into the cart and under the blankets. I then pulled them back along the towpath to the cottage, where there was a fire burning in the stove waiting for them, and hot-water bottles in beds with freshly washed and ironed sheets. This had taken me all afternoon, paying particular attention to the many spiders who have taken up residence in the cottage, as my eldest granddaughter Iris suffers from mild arachnophobia. The girls called out excitedly when they saw the occasional bat flying overhead, and pointed to the stars, which they could see much more clearly than I could. We passed the narrowboats moored beside the towpath, a few with lit portholes, and figures seen dimly inside. There was a low mist over the canal.

I remembered my mother telling me of one of her clearest childhood memories – of also being pulled along in a cart at night, wrapped in a fur coat in autumn, beneath a sky full of stars in rural Germany in the early 1920s. Her mother was taking her to the railway station near Biere, a small village in the Altmark, south-west of Berlin, after staying in her grandparents' house. Her grandfather was the local doctor and my mother had also told me stories of her fascination at occasionally seeing him at work. Reducing a dislocated shoulder by putting an empty bottle in the patient's armpit and then pulling the arm; dealing with a severed artery pumping blood in a farmworker with a scythe injury. Perhaps these stories influenced me in my eventual decision to become a doctor. She was certainly very pleased with my choice, although I had caused her and my father much grief in the years before I made it.

I have a photograph on my kitchen wall of my mother at the age of eleven, with her elder sister and younger brother, in Magdeburg in 1929. It is beautifully done. I look at it every day.

They are looking directly at the camera, so that their questioning eyes, in black and white, follow you around the room. My mother and her sister Sabine are wearing simple white blouses, their young brother Hans Marquadt is wearing a sailor suit. This was six years after the German hyperinflation in which her parents had lost most of their modest wealth. The Great Depression had started, and the Nazis were on the rise. They could not have had any idea what the future held for them. Sabine

was to become an enthusiastic Nazi, my uncle a Luftwaffe fighter pilot in the crack Schlageter 26 Squadron, and my mother a dissident, who fled to England in 1939, having been denounced to the Gestapo. Sabine was killed in a British air raid on Jena in 1945 and Hans Marquadt was shot down over Kent in 1940. He survived the war as a POW but never married and died from alcoholism in 1967.

My mother wrote a memoir of her life, working slowly at it over the years, which my brother and elder sister edited and had published privately after her death. It is written in perfect English. I reread it recently – to my shame I hadn't read it properly before. It describes quite devastating loss, of her family, of her past, of her culture, of her childhood – all destroyed by the Nazis and war. She tells us of how there was a very strict rule in her family – a very close and loving family – that one should not make a fuss. *Stell dich nicht so an.* It is a rule she applied to a certain extent with her own four children in England, though with singular failure in my case. She unfailingly adhered to it in her writing, so she describes the destruction of her past in the most restrained and understated tones – almost infuriatingly so, as you can sense so much passion beneath it all.

Reading the book filled me with me an intense longing to talk to her again – and not just to reassure her that I have been successful in ways of which she would approve. It also filled me with deep sadness that I was so preoccupied with my own life that I took little real interest in hers while she was still alive. In the book, she writes of how when she was interrogated by the Gestapo, she denied belonging to the anti-Nazi Christian Confessing Church. She had torn up her membership card the day before as a precaution. This left her, she tells us, with the feeling that she had betrayed both her faith and her two colleagues with whom she had been denounced. They were tried and sentenced to prison, while she was able to flee to England before their trial, where she was to have been a witness. And yet I had never really appreciated during her life how much she suffered from the guilt of a survivor, and I never discussed this with her.

Why is it that only in old age, and closer to death, I have come to understand so much more about myself and my past? We are like little boats that our parents launch onto the ocean, and we sail round the world, full circle, to return finally to the harbour from which we started, but by then our parents are long gone.

My mother loved her grandparents and especially their garden, which she described as a paradise. Seventy years later, shortly before her death at the age of eighty-two from breast cancer – though we didn't know at the time that she was already dying – my brother and I took her back to her home town of Magdeburg, and to Biere. Magdeburg had been largely destroyed by a single air raid on 16 January 1945. There was no trace of her family home or even of the street on which it had stood. We drove to Biere on the road along which she had been pulled in the cart seventy-five years earlier. She pointed out to us how the road was quite unchanged – with a cinder track beside the asphalt for horses and wagons.

Her grandparents' house was still standing, though the garden was overgrown and neglected. Her grandparents had been buried in the village churchyard, but there is a law in Germany that you are only entitled to a limited number of years' rest in a cemetery and, to her disappointment, when we went to look, their headstones had disappeared.

In the first few weeks after I was diagnosed with advanced cancer and when I did not know if I had disseminated disease or not and might soon be dead, I would sit at the kitchen table and look at the photograph. Physicists talk of 'block time' – that the past, present and future are all equally real. In Einstein's equations for relativity and space-time, time can run forwards or backwards. There is nothing inevitable about time always moving irredeemably forwards. The arrow of time that dominates our lives, ages our bodies and ultimately kills us, has no place, apparently, in theoretical physics. The present is a place, and past and future are simply other places, just as the place I am in at the moment is one place among many others on the surface of the planet. Looking into my mother's young eyes, my own life now possibly nearing its end, I felt as close as I could ever possibly be to living in block time – past, present and future all combined in one whole.

Postscript

Six months after the radiotherapy finished, my PSA was measured again. Although I knew it was most unlikely that the cancer would already be growing back, I was anxious about the possible result for weeks before the blood test and could think of little else. I had been told I would receive a phone call at 11 o'clock in the morning but had to wait two hours before my phone eventually rang. I was told that my PSA had fallen to zero point one. This is as low as it can get. It does not mean that I am cured (as family and friends all like to think) – my very high presenting PSA comes with a 75 per cent probability of recurrence within the next five years. But chemotherapy could then be used, so I will probably live for a few years yet, although it can never be certain. This uncertainty produces a little flash of anxiety. But what, I then ask myself irritably, do you want? To live for ever? To become old and decrepit? And once again I marvel at my ridiculous inability to accept the inevitability of my death – indeed, its necessity.

The phone call brought great relief, and I was filled with mistaken optimism that my life would return to how it had been before the diagnosis of cancer. After a year of chemical castration, the side effects were becoming quite trying – mainly of fatigue and muscular weakness, which I found offensive. Perhaps, I told myself, my lack of energy was the product of anxiety and would now improve. At least I would not need to worry about the cancer coming back until the next PSA test in six months' time. I would be free for a while.

But any relief did not last long, as ten days later Putin invaded Ukraine. I returned to a state of chronic anxiety and pre-occupation, my cancer now entirely forgotten.

When I first went to Ukraine thirty years ago, I encountered a medical system that reflected Soviet society in miniature: it was ruled over by totalitarian professors who would brook no dissent. I saw my role as political as well as clinical – I was helping young doctors rebel against the monolithic hierarchy that ruled over them. In retrospect, I was naive and misunderstood much of what I saw, and probably contributed very little to medicine in Ukraine. But I made many very close friends and came to see the country as my second home. Ukraine has been struggling since independence in 1991 to escape its past under the Russian yoke. The freedom it now enjoys would be a fatal threat to Putin's despotic kleptocracy if it spread to Russia. Putin would rather have his soldiers commit mass murder and perpetrate atrocities than let this happen.

At the time of writing, in spring 2022, it is impossible to know what will happen in Ukraine, other than that the country will be devastated, with many thousands of people killed and millions displaced. But the Ukrainians will fight to the death. I always knew they would. They see no alternative.

I phone my friends in Lviv and Kyiv every day. Sometimes I can hear the air-raid sirens in the background. I know as much about the course of the war from the media as they do, so we ironically discuss the weather alongside the war crimes that Putin and his soldiers are perpetrating. I find the contrast between their lives and mine difficult, but I think they like to hear my voice, and to know that there is still a more peaceful, civilised world beyond the nightmare in which they are now living, and that so many people all over the world are concerned for their fate. My friends' lives have been utterly changed, just as my mother's was. I never dreamed that I would live to see history repeat itself, in such horror. I do not know if I will ever see Ukraine again, or even ever see my friends again. But we have a duty to be optimistic – if we are not, and we give up, then evil will certainly triumph. I will return.

Acknowledgements

Many friends looked at earlier drafts of this book, and all made very helpful comments. I would like to thank Robert McCrum, JP Davidson, Erica Wagner, Sarah Marsh, Rachel Clarke, David Fickling, Gina Cowen, John Milnes, Paul Klemperer, Pedro Ferreira and Roman Zoltowski.

Once again, I am indebted to my editor Bea Hemming who has played a very major role in sorting out the muddle with which I presented her. The support from my agent Julian Alexander has been second to none. The book would never have seen the light of day without the love and encouragement of my wife Kate.

Patrick Sherlock

Henry Marsh is a retired neurosurgeon and the author of the *New York Times* bestselling memoir *Do No Harm* and National Book Critics Circle Award finalist *Admissions.* He has been the subject of two documentary films, *Your Life in Their Hands,* which won the Royal Television Society Gold Medal, and *The English Surgeon,* which won an Emmy. He lives in London.